Library of
Davidson College

CONCEPTS
OF
RADIATION DOSIMETRY

PERGAMON TITLES OF RELATED INTEREST

Cember, H. — *Introduction to Health Physics*
Gollnick, D.A. — *Experimental Radiological Health Physics*

CONCEPTS
OF
RADIATION DOSIMETRY

Kenneth R. Kase
Department of Radiation Therapy
Joint Center for Radiation Therapy
Harvard Medical School

and

Walter R. Nelson
Radiation Physics Group
Stanford Linear Accelerator Center
Stanford University

PERGAMON PRESS
New York / Toronto / Oxford / Sydney / Frankfurt / Paris

Pergamon Press Offices:

U.S.A.	Pergamon Press Inc., Maxwell House, Fairview Park, Elmsford, New York 10523, U.S.A.
U.K.	Pergamon Press Ltd., Headington Hill Hall, Oxford OX3, OBW, England
CANADA	Pergamon of Canada, Ltd., 75 The East Mall, Toronto, Ontario M8Z 5W3, Canada
AUSTRALIA	Pergamon Press (Aust) Pty. Ltd., 19a Boundary Street, Rushcutters Bay, N.S.W. 2011, Australia
FRANCE	Pergamon Press SARL, 24 rue des Ecoles, 75240 Paris, Cedex 05, France
WEST GERMANY	Pergamon Press GmbH, 6242 Kronberg/Taunus, Frankfurt-am-Main, West Germany

Copyright (c) 1978
Pergamon Press Inc.

Library of Congress Cataloging in Publication Data

Kase, Kenneth R   1938-
   Concepts of radiation dosimetry.

   "Developed from a set of notes which accompanied a seminar series . . . given by the authors at Stanford University during the spring quarter, 1970."
     Bibliography: p.
     Includes index.
   1. Radiation dosimetry.   I. Nelson, Walter Ralph, 1937-   joint author. II. Title.
(DNLM: 1. Radiometry. WN650.3 K19c)
QC795.32.R3K37   1978 539.7'7   78-5705
ISBN 0-08-023162-4
ISBN 0-08-023161-6 pbk.

All Rights Reserved. No part of this publication may be reproduced, stored in a retrieval system or transmitted in any form or by any means: electronic, electrostatic, magnetic tape, mechanical, photocopying, recording or otherwise, without permission in writing from the publishers.

Printed in the United States of America

CONTENTS

			Page
Chapter 1	Basic Concepts		1
	1.1	Introduction	1
	1.2	Dosimetry Terminology	1
	1.3	Stochastic and Macroscopic Quantities	4
	1.4	Exposure	7
	1.5	Energy Imparted and Energy Transferred	7
	1.6	Charged Particle Equilibrium	12
		References	13
Chapter 2	The Interaction of Electromagnetic Radiation with Matter		14
	2.1	Introduction	14
	2.2	Negligible Processes	16
	2.3	Minor Processes	18
	2.4	Major Processes	19
	2.5	Attenuation and Absorption	31
		References	33
Chapter 3	Charged Particle Interactions		34
	3.1	Introduction	34
	3.2	Kinematics of the Collision Process	35
	3.3	Collision Probabilities with Free Electrons	38
	3.4	Ionization Loss	43
	3.5	Restricted Stopping Power	47
	3.6	Compounds	48
	3.7	Gaussian Fluctuations in the Energy Loss by Collision	48
	3.8	Landau Fluctuations in the Energy Loss by Collision	53

			Page
	3.9	Radiative Processes and Probabilities	53
	3.10	Radiative Energy Loss and the Radiation Length	58
	3.11	Comparison of Collision and Radiative Energy Losses for Electrons	59
	3.12	Radiation Energy Losses by Heavy Particles	61
	3.13	Fluctuations in the Energy Loss by Radiation	64
	3.14	Range and Range Straggling	64
	3.15	Elastic Scattering of Charged Particles	67
	3.16	Scaling Laws for Stopping Power and Range	78
		References	83
Chapter 4		Energy Distribution in Matter	85
	4.1	Introduction	85
	4.2	Linear Energy Transfer	86
	4.3	Delta Rays	88
	4.4	LET Distributions	91
	4.5	Event Size	93
	4.6	Local Energy Density	94
	4.7	Conclusions	99
		References	102
Chapter 5		Dose Calculations	103
	5.1	Introduction	103
	5.2	Sources	103
	5.3	Flux Density	104
	5.4	Point Isotropic Source	105
	5.5	Line Source	106
	5.6	Area Source	111

		Page
5.7	Infinite Slab Source	115
5.8	Right-Circular Cylinder Source: Infinite-Slab Shield, Uniform Activity Distribution	118
5.9	Spherical Source: Infinite-Slab Shield, Uniform Activity Distribution	122
5.10	Spherical Source: Field Position at Center of Sphere	125
5.11	Transport of Radiation	128
5.12	Buildup Factor Corrections to the Uncollided-Flux Density Calculations	131
5.13	Approximating the Buildup Factor with Formulas	135
5.14	Calculation of Absorbed Dose from Gamma Radiation	137
	References	144

Chapter 6 **Measurement of Radiation Dose—Cavity-Chamber Theory**   145

6.1	Introduction	145
6.2	Cavity Size Small Relative to Range of Electrons	146
6.3	The Effect of Cavity Size	153
6.4	Measurement of Absorbed Dose	158
6.5	Applications of Cavity Theory for Photon Fields (f and C_λ)	163
6.6	Applications of Cavity Theory for Electron Fields (C_E)	166
6.7	Average Energy Associated with the Formation of One Ion Pair (W)	168
	References	169
	Appendix	171
	Subject Index	208

LIST OF FIGURES

Figure	Caption	Page
1.1	Energy density as a function of the mass.	6
1.2	Energy imparted with CPE condition.	8
1.3	Case where $E_D \neq E_K$ even though CPE exists.	11
2.1	Pair production in the field of a nucleus. Compton interaction. Bremsstrahlung.	27
2.2	An incident photon (no track) undergoes pair production in the field of an electron (triplet).	28
3.1	An incident photon (no track) undergoes a pair production interaction.	39
3.2	Bremsstrahlung.	57
3.3	Electron fractional energy loss per radiation length.	62
3.4	Number-distance curve for charged particles.	65
3.5	Mass stopping power for heavy charged particles.	82
4.1	The ratio $P = LET_\Delta / LET_\infty$ of electrons (a) and positrons (b) as a function of $g = m\Delta/T$.	90
4.2	Maximum energy density for protons and electrons traversing a 1 μm diameter sphere of tissue.	95
4.3	Probability of increment ΔZ in 7 and 1 μm spheres. $P'(\Delta A) = P(\Delta Z)/\Delta Z$	100
4.4	Local energy density frequencies in 7 and 1 μm spheres. Dose = 7.5 rads.	100
4.5	Local energy density frequencies in 7 and 1 μm spheres. Dose = 75 rads.	101
4.6	Local energy density frequencies in 7 and 1 μm spheres. Dose = 750 rads.	101

5.1	Radiation at various distances from a right circular cylindrical source.	123
5.2	Radiation at various distances from a hemispherical source.	127
6.1	Stopping power ratios in carbon and lead using various theoretical treatments.	154
6.2	Electron distributions in various size cavities	155
A.1-A.13	The exponential integrals E_1 and E_2 along with e^{-x}.	172-184
A.14-A.19	The Sievert integrals (F functions).	185-190
A.20-A.24	The parameters necessary for determining self-absorption in cylindrical and spherical sources.	191-195
A.25-A.30	Buildup factors in lead, iron and water.	196-201
A.31-A.36	Parameters required for calculating buildup factors in iron, water, lead and concrete.	202-207

PREFACE

This monograph is developed from a set of notes which accompanied a seminar series on the Concepts of Radiation Dosimetry given by the authors at Stanford University during the Spring Quarter 1970. The manuscript discusses the basic information required to understand the principles of photon and charged particle dose measurement from basic particle interactions to cavity chamber theory. As health physicists at the Stanford Linear Accelerator Center we were interested in the dosimetry of high energy photons and charged particles. Thus, throughout the text we have emphasized the extension of dosimetry principles to the high energy situation. We hope that the reader will gain some insight to the dosimetry of particles such as pions and muons as well as high energy electrons and photons. Because the audience during the seminars was composed primarily of experienced health physicists, radiological physicists, nuclear engineers, and medical doctors, the material is presented at a level requiring advanced understanding of mathematics and physics.

A detailed development of all the theories involved is not included because these have been adequately covered in several texts. We have attempted to discuss the pertinent theories and their relationship to dosimetry. What we have tried to do is gather together in one place the information necessary for charged particle and photon dosimetry, citing appropriate references the reader may consult for further background or a more complete theoretical treatment. We hope this monograph will be useful to the health physicist and radiological physicist.

The material in this monograph was drawn primarily from the following references:

1. F. H. Attix, W. C. Roesch, and E. Tochilin, Radiation Dosimetry, Second Edition, Volume I, Fundamentals (Academic Press, New York, 1968).
2. J. J. Fitzgerald, G. L. Brownell, and F. J. Mahoney, Mathematical Theory of Radiation Dosimetry (Gordon and Breach, New York, 1967).

3. K. Z. Morgan and J. E. Turner, <u>Principles of Radiation Protection</u> (Krieger Publishing Co., New York, 1973).

In the text, direct reference to these books will be made using the notation (ART), (FBM) and (MT). Additional references are cited at the end of each chapter and will be indicated in the text by number.

The authors gratefully acknowledge the encouragement and support of Dr. Richard McCall and Wade Patterson and in particular Professors C. J. Karzmark (Radiology) and T. J. Connolly (Nuclear Engineering) of Stanford University for sponsoring the seminar. We thank Dr. H. DeStaebler for reviewing Chapters 2 and 3 and Dr. Goran Svensson for reviewing Chapter 6. In general, their criticism has been very helpful to us. The bubble chamber pictures were provided by Dr. James Loos of Experimental Group B at SLAC, and were prepared by G. Fritzke. Finally, we thank the 40 or so people who attended the seminars and contributed to the discussion.

As this book reaches the final stages before publication, we wish to thank our many colleagues who have used the first set of notes and have provided us with useful criticism for updating the manuscript. In particular, the support from Ron Kathren, whose encouragement was instrumental in publishing this book, is truly appreciated.

Harvard UniversityKenneth R. Kase
Stanford UniversityWalter R. Nelson
December 1977

CHAPTER 1

BASIC CONCEPTS

1.1 Introduction

Before embarking on a study of radiation dosimetry it is necessary to understand the basic concepts and terminology involved. The history of radiation dosimetry is fraught with many, sometimes confusing, concepts and definitions. We will discuss dosimetry using the concepts, quantities and units defined by the International Commission on Radiological Units and Measurements (ICRU) in their 1971 Report 19, "Radiation Quantities and Units."[1] The definitions used in this monograph taken from ICRU Report 19 are presented in Section 1.2. Following the definitions we discuss some of the basic concepts involved in the quantities defined.

1.2 Dosimetry Terminology

1. Directly Ionizing Particles — charged particles having sufficient kinetic energy to produce ionization by collision.

2. Indirectly Ionizing Particles - uncharged particles which can liberate directly ionizing particles or can initiate nuclear transformations.

3. Exposure (X) - the quotient of dQ by dm where dQ is the absolute value of the total charge of the ions of one sign produced in air when all the electrons liberated by photons in a volume element of air whose mass is dm are completely stopped in air.

$$X = dQ/dm$$

The special unit of exposure is the <u>roentgen</u> (R).

$$1 R = 2.58 \times 10^{-4} \, C \, \text{-kg}^{-1}$$

4. Absorbed Dose (D) - the quotient of $d\overline{E}_D$ by dm where $d\overline{E}_D$ is the mean energy imparted by ionizing radiation to the mass dm of matter in a volume element.

$$D = d\overline{E}_D/dm$$

The special unit of absorbed dose is the Gray (Gy).

$$1 \text{ Gy} = 1 \text{ J-kg}^{-1} \quad (= 100 \text{ rad})$$

5. Energy Imparted (E_D) - a stochastic quantity which is the difference between the sum of the kinetic energies of all the directly and indirectly ionizing particles which have entered a volume ($\sum E_E$) and the sum of the kinetic energies of all those which have left it ($\sum E_L$) minus the energy equivalent of any increase in rest mass ($\sum E_R$) that took place in nuclear or elementary particle reactions within the volume.

$$E_D = \sum E_E - \sum E_L - \sum E_R$$

6. Mean Energy Imparted ($\bar{E}_D$) - the expectation value of the energy imparted, sometimes referred to as integral dose.

7. Mean energy expended in a gas per ion pair formed ($\overline{W}$) - the quotient of E by N, where N is the number of ion pairs formed when a directly ionizing particle of initial kinetic energy E is completely stopped by the gas.

$$\overline{W} = E/N$$

8. Particle Fluence (Φ) - the quotient of dN by da where dN is the number of particles which enter a sphere of cross sectional area da.

$$\Phi = dN/da$$

9. Particle Flux Density (ϕ) - the quotient of $d\Phi$ by dt where $d\Phi$ is the particle fluence in time dt.

$$\phi = d\Phi/dt$$

10. Energy Fluence (F) — the quotient of dE_f by da where dE_f is the sum of the energies, exclusive of rest energies, of all the particles which enter a sphere of cross sectional area da.

$$F = dE_f/da$$

11. Energy Flux Density (I) — the quotient of dF by dt where dF is the energy fluence in the time dt.

$$I = dF/dt$$

12. Kerma (K) — the quotient of dE_K by dm where dE_K is the sum of the initial kinetic energies of all the charged particles liberated by indirectly ionizing particles in a volume element of the specified material. dm is the mass of the matter in that volume element.

$$K = dE_K/dm$$

13. Mass Attenuation Coefficient (μ/ρ) — for a given material, μ/ρ for indirectly ionizing particles is the quotient of dN by the product of ρ, N and dl where N is the number of particles incident normally upon a layer of thickness dl and density ρ, and dN is the number of particles that experience interaction in this layer.

$$\mu/\rho = \frac{1}{\rho N} \frac{dN}{dl}$$

14. Mass Energy Transfer Coefficient (μ_K/ρ) — for a given material, μ_K/ρ for indirectly ionizing particles is the quotient of dE_K by the product of E, ρ and dl where E is the sum of the energies (excluding rest energies) of the indirectly ionizing particles incident normally upon a layer of thickness dl and density ρ, dE_K is the sum of the kinetic energies of all the charged particles liberated in this layer.

$$\mu_K/\rho = \frac{1}{E\rho} \frac{dE_K}{dl}$$

15. Mass Energy Absorption Coefficient (μ_{en}/ρ) — for a given material, μ_{en}/ρ for indirectly ionizing particles is (μ_K/ρ) (1 - G) where G is the proportion of the energy of secondary charged particles that is lost to bremsstrahlung in the material.

16. Mass Stopping Power $(S/\rho)^*$ — for a given material, S/ρ for charged particles is the quotient of dE_s by the product of ρ and dl where dE_s is the average energy lost by a charged particle of specified energy in traversing a path length dl, and ρ is the density of the medium.

$$S/\rho = \frac{1}{\rho} \frac{dE_s}{dl}$$

17. Linear Energy Transfer (LET)* — for charged particles in medium, LET is the quotient of dE_L by dl where dE_L is the average energy locally imparted to the medium by a charged particle of specified energy traversing a distance dl.

18. Charged Particle Equilibrium (CPE) — CPE exists at a point P centered in a volume V if each charged particle carrying a certain energy out of V is replaced by another identical charged particle which carries the same energy into V. If CPE exists at a point then D = K at that point provided that bremsstrahlung production by secondary charged particles is negligible.

1.3 Stochastic and Macroscopic Quantities

Many of the quantities defined are macroscopic quantities such as absorbed dose, exposure, fluence, etc. On the other hand, stochastic quantities such as energy imparted, charge liberated, etc., may vary greatly from point to point since radiation fields are in general not uniform in space. Consequently, these quantities must be

---

*A discussion of these terms is given in Chapter 3.

determined for sufficiently small regions of space or time by some limiting procedure. We illustrate this procedure using the quantity "absorbed dose."

Absorbed dose is a measure of energy imparted to a medium divided by the mass of the medium. If we choose a large mass element and measure the energy imparted, we will obtain a value of $E/m)_1$ (see Fig. 1.1). Now, if we take a smaller mass element and measure the value $E/m)_2$, in general we find $E/m)_2$ will be larger than $E/m)_1$. When m is large enough to cause significant attenuation of the primary radiation (e.g., x rays), the fluence of charged particles in the mass element under consideration is not uniform. This causes the ratio E/m to increase as the size of the mass m is decreased.

As m is further reduced we will find a region in which the charged particle fluence is sufficiently uniform that the ratio E/m will be constant. It is in this region that the ratio E/m represents absorbed dose. Thus, the expectation value $\bar{E}$ of the energy imparted over an appropriate size mass element must be used to determine absorbed dose.

At the other extreme, m must not be so small that the energy deposition is caused by a few interactions. If m is further decreased from the region of constant E/m, we will find that the ratio will diverge. That is, as m gets very small the energy deposition is determined by whether or not a charged particle interacts within m. Consequently, E will be zero for many mass elements and very large for others. These fluctuations occur because charged particles lose energy in discrete steps. Hence, the determination of absorbed dose also requires that the mass element m be large enough so that the energy deposition is caused by many particles and many interactions.

Similar discussions may be made for other quantities and it must be realized that the macroscopic quantities defined using the differential notation imply that a limiting process as described above has occurred.

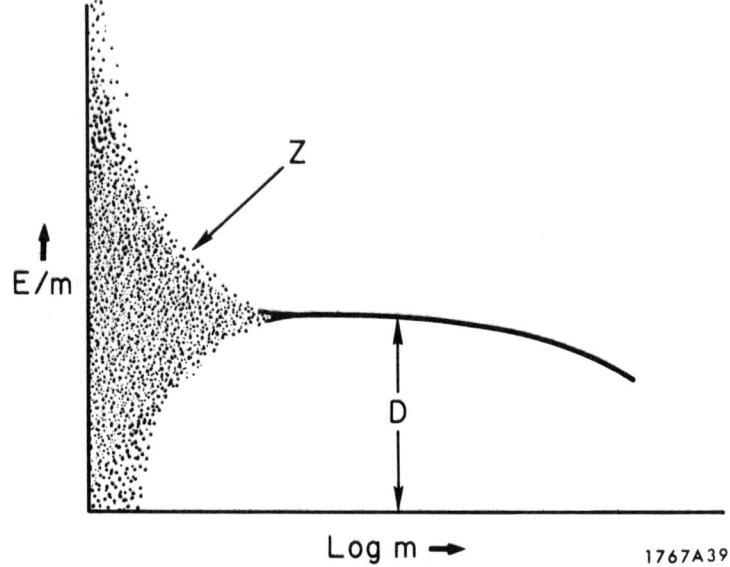

FIG. 1.1

Energy density as a function of the mass for which energy density is determined. The horizontal line covers the region in which the absorbed dose can be established in a single measurement. The shaded portion represents the range where statistical fluctuations are important. (From (ART), Chapter 2.)

1.4 Exposure

The quantity, exposure, as currently defined requires that all the electrons liberated by photons in a mass element of air be completely stopped in air. It also requires that all the ions (of one sign) produced by these electrons be collected. To make any absolute measurement of exposure, therefore, requires use of a free air ionization chamber. This in turn puts an upper limit on the photon energy for which absolute exposure measurements are practicable. This energy limit (a few hundred KeV) is determined by the range of the electrons and the ion chamber size.

In principle there is no energy limit on the quantity dQ/dm. There is simply a practical limit on the accuracy with which exposure can be measured as the photon energy increases. Relative measurement of exposure can be made at any photon energy using air-equivalent cavity chambers (see Chapter 6). The accuracy of these measurements depends on the photon energy and the chamber construction. Accuacies of 1-2% can be achieved for photons up to a few MeV. As the photon energy increases, the uncertainty in the measurement increases because of failure to collect all the ions produced by electrons liberated in the mass element. Further uncertainty is introduced when there is significant attenuation of the photon field within the range of the electrons liberated by those photons. Consequently, the quantity exposure as presently defined is practical only for photon fields below a few MeV in energy.

1.5 Energy Imparted and Energy Transferred (Absorbed Dose and Kerma)

To better understand absorbed dose, kerma and charged particle equilibrium, one must understand how the energy balance is made for a mass element exposed to radiation. Figure 1.2 is a schematic drawing showing 10 photons incident on a mass element. Each in some way involves the movement of energy into and out of the mass. Table 1.1 gives an arbitrary breakdown of the energy entering and leaving the mass on charged and uncharged particles.

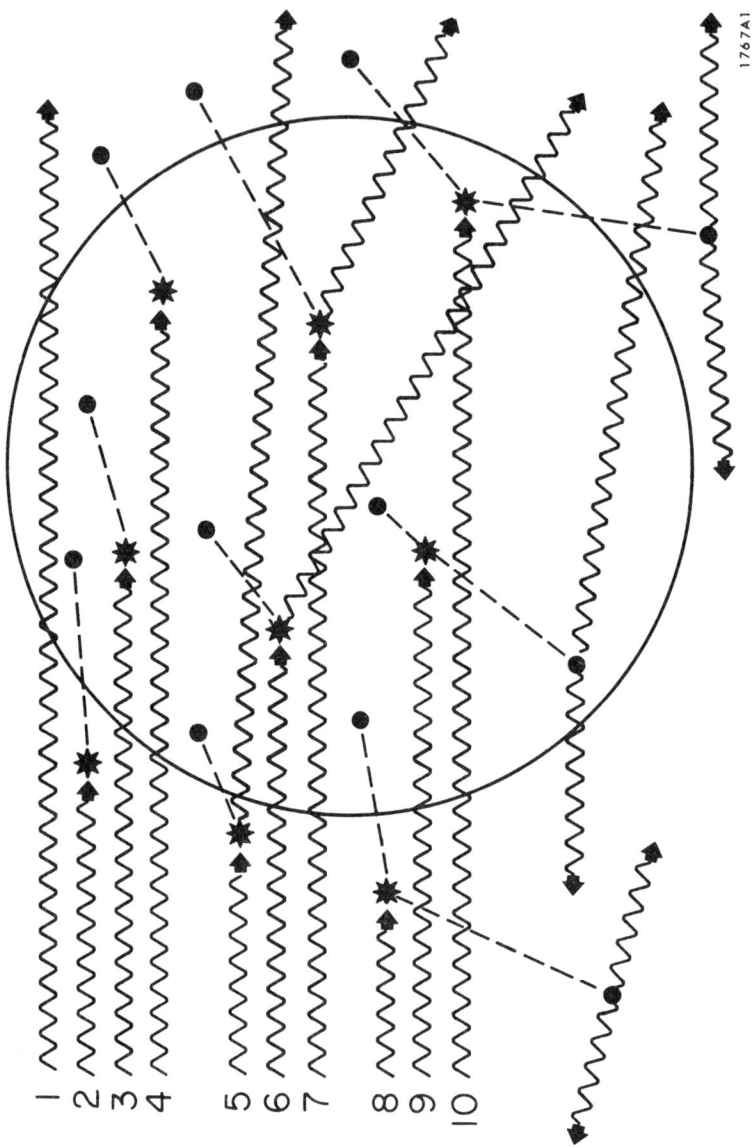

FIG. 1.2 Energy imparted with CPE condition.

TABLE 1.1

	Primary γ Energy	Secondary γ Energy	Secondary Charged Particle Energy		$(E_E)_c$	$(E_L)_c$	$(E_E)_u$	$(E_L)_u$	$(E_R)_u$
			e^-	e^+					
1	.5	-	-	-	0	0	.5	.5	0
2	.5	-	.5	-	.3	0	0	0	0
3	.5	-	.5	-	0	0	.5	0	0
4	.5	-	.5	-	0	.2	.5	0	0
5	1.0	.5	.5	-	.3	0	.5	.5	0
6	1.0	.5	.5	-	0	0	1.0	.5	0
7	1.0	.5	.5	-	0	.2	1.0	.5	0
8	3.0	-	1.0	1.0	.8	0	0	0	0
9	3.0	-	1.0	1.0	0	0	3.0	1.0	0
10	3.0	-	1.0	1.0	0	1.0	3.0	0	1.0
Σ					1.4	1.4	10.0	3.0	1.0

The energy entering and leaving the mass on charged particles is denoted by $(E_E)_c$ and $(E_L)_c$ respectively; the energy entering and leaving on uncharged particles is denoted by $(E_E)_u$ and $(E_L)_u$ respectively; while $(E_R)_u$ denotes the energy which goes into the creation of rest mass within the mass element. The energy imparted to the mass element (E_D) is equal to the algebraic sum of all the energy components,

$$E_D = (\sum E_E)_c - (\sum E_L)_c + (\sum E_E)_u - (\sum E_L)_u - (\sum E_R)_u$$

This is the energy used to calculate absorbed dose and for this example it is

$$E_D = 1.4 - 1.4 + 10.0 - 3.0 - 1.0 = 6.0 \text{ MeV}$$

If none of the charged particles radiate energy within the mass, the energy transferred to charged particles in the mass element (E_K) is determined by the algebraic sum of the uncharged particle energy terms and in this example is:

$$E_K = 10.0 - 3.0 - 1.0 = 6.0 \text{ MeV}$$

This is the energy used to calculate kerma.

In this example, the energy entering the mass element on charged particles is exactly balanced by energy leaving on charged particles, i.e.,

$$(\textstyle\sum E_E)_c - (\textstyle\sum E_L)_c = 1.4 - 1.4 = 0$$

Thus, we say charged particle equilibrium (CPE) exists. Also, since none of the secondary charged particles produce bremsstrahlung within the mass element, $E_D = E_K$, and consequently the absorbed dose will equal the kerma.

When the secondary charged particles lose energy by bremsstrahlung production within the mass element, absorbed dose and kerma will not be equal even though CPE exists. This situation is illustrated in Fig. 1.3. In this case, we assume that $(\textstyle\sum E_E)_c - (\textstyle\sum E_L)_c = 0$ and that there is no energy lost in rest mass increases $(\textstyle\sum E_R)_u = 0$. Consequently the energy imparted to the mass is:

$$E_D = (\textstyle\sum E_E)_{u_0} - (\textstyle\sum E_L)_{u_1} - (\textstyle\sum E_L)_{u_2}$$

Whereas the energy transferred to charged particles by uncharged particles within the mass element is

$$E_K = (\textstyle\sum E_E)_{u_0} - (\textstyle\sum E_L)_{u_1}$$

obviously $E_D \neq E_K$ and so absorbed dose will not equal kerma in this case. This occurs because in E_K we consider only the energy transferred to charged particles in the mass element and do not consider how the charged particles subsequently lose their energy. Energy imparted (E_D) on the other hand is a total energy balance considering charged and uncharged particles.

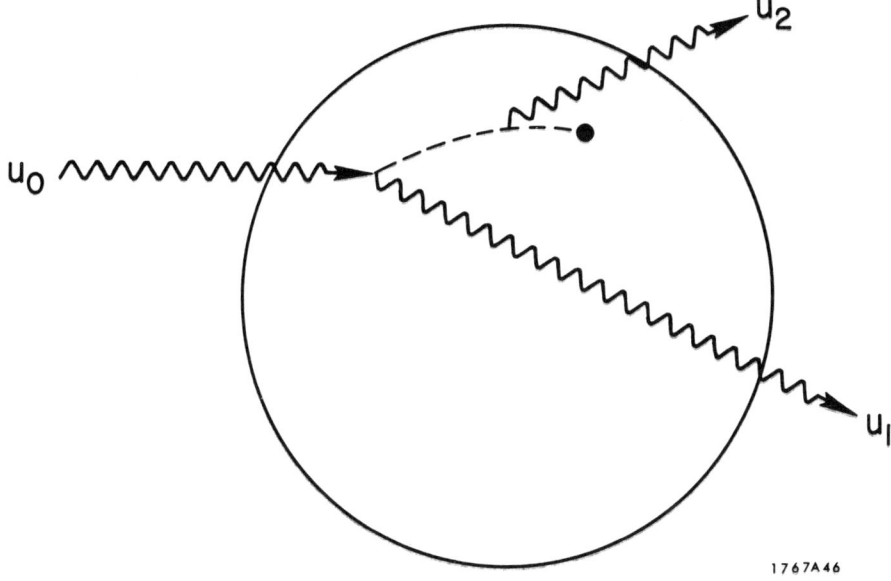

FIG. 1.3

Case where $E_D \neq E_K$ even though CPE exists.

1.6 Charged Particle Equilibrium

The concept of charged particle equilibrium deserves a short discussion. If each charged particle carrying a certain energy out of a mass element is replaced by another identical charged particle carrying the same energy in, then CPE is said to exist in the mass element. This does not necessarily require that the number of charged particles entering be equal to the number leaving. It does require that the energy entering on charged particles equal the energy leaving on charged particles.

CPE will generally exist in a uniform medium at points which lie more than the maximum range for the secondary charged particles from the boundaries of the medium. CPE will generally not exist near the interface between two dissimilar media. For purposes of absorbed dose measurement CPE is not necessary as long as the appropriate corrections are made. We will discuss this in more detail in Chapter 6.

REFERENCES

1. ICRU Report 19, Radiation Quantities and Units, International Commission on Radiation Units and Measurements, Washington, D.C. (1971).

(ART) F. H. Attix, W. C. Roesch, and E. Tochilin (eds.), Radiation Dosimetry, Second Edition, Volume I, Fundamentals (Academic Press, New York, 1968).

CHAPTER 2

THE INTERACTION OF ELECTROMAGNETIC RADIATION WITH MATTER

2.1 Introduction

Essentially, there are twelve possible processes by which the electromagnetic field of a photon may interact with matter.[1] These are classified in Table 2.1,[2] where the major processes are "boxed in," the minor processes ($\geq 1\%$ contribution over certain energy intervals) are "underlined," and the rest are negligible processes (note that some processes have been completely omitted because of their rare occurrence).

The symbols τ, r, and κ refer to cross sections (or coefficients) of the various interaction processes. The units of these cross sections can be barns/atom, cm^2/g or cm^{-1} and the appropriate units will be clear from the context. The following equations illustrate the conversion from one set of units to another

$$\tau(cm^2/g) = \tau(b/atom) \frac{N_0}{A} \times 10^{-24} \quad \text{(usually written } \tau/\rho) \quad (2.1)$$

$$\tau(cm^{-1}) = \tau(b/atom) \frac{N_0}{A} \rho \times 10^{-24} \quad (2.2)$$

Also

$$\tau_{pe} = \tau_K + \tau_L + \cdots$$

$$\sigma_{pn} = \sigma(\gamma, n) + \sigma(\gamma, p) + \sigma(\gamma, f) + \cdots$$

are total cross sections for the atomic and nuclear photo effects, respectively.

Elastic scattering refers to the fact that kinetic energy is conserved in the process. When inelastic scattering occurs, kinetic energy is not conserved. For example, in the case of Compton scattering, some of the energy is needed to overcome the binding energy of the electron to the atom. The rest appears as kinetic energy of the photon and electron. If the individual scattering elements (such as electrons

TABLE 2.1

CLASSIFICATION OF PHOTON INTERACTIONS

Type of Interaction Interaction with	ABSORPTION A	SCATTERING — ELASTIC (Coherent) B	SCATTERING — INELASTIC (Incoherent) C
I ATOMIC ELECTRONS	Photoelectric Effect $\tau_{pe} \begin{cases} \sim Z^4 \text{ (low energy)} \\ \sim Z^5 \text{ (high energy)} \end{cases}$	Rayleigh Scattering $\sigma_R \sim Z^2$ (low energy limit)	Compton Scattering $\sigma \sim Z$
II NUCLEONS	Photonuclear Reactions $(\gamma, n), (\gamma, p), (\gamma, f)$, etc. $\sigma_{pn} \sim Z$ $(h\nu \gtrsim 10 \text{ MeV})$	Elastic Nuclear Scattering	Nuclear Resonance Scattering
III ELECTRIC FIELD OF SURROUNDING CHARGED PARTICLES	Pair Production a. Field of Nucleus $\kappa_n \sim Z^2 (h\nu \geq 1.02 \text{ MeV})$ b. Field of Electron $\kappa_e \sim Z (h\nu \gtrsim 2.04 \text{ MeV})$	Delbruck Scattering	
IV MESONS	Photomeson Production $h\nu \gtrsim 140 \text{ MeV}$		

or nucleons) are virtually free, they scatter independently of one another — thus the term <u>incoherent</u> scattering. Complementary to this, one refers to <u>coherent</u> scattering as a type of scattering in which the individual scattering elements act as a whole. Incoherent scattering implies inelastic scattering. Coherent scattering implies elastic scattering.

2.2 Negligible Processes

A. Elastic Nuclear Scattering (II-B)

This is regarded as the nuclear analog to very low energy Compton scattering by an electron. This seems inconsistent since Compton scattering is an inelastic process whereas elastic nuclear scattering is in the "elastic" category! A digression into Compton scattering is in order at this point.

First of all, Compton scattering is described (quantum mechanically) by the Klein-Nishina differential scattering cross section, which reduces to

$$\frac{d\sigma}{d\Omega} = \frac{Ze^4}{2m^2c^4}(1 + \cos^2\theta)(cm^2/atom\text{-}sr) \qquad (2.3)$$

where

$$mc^2 = \text{electron rest mass}$$

$$\theta = \text{angle of scattered photon}$$

in the limit as $h\nu \longrightarrow 0$! But this is equivalent to a classical result obtained by Thomson,[3] who treated the process as an elastic one in which the free electron vibrates under the influence of the photon's electric field, and re-emits photon radiation of the same frequency (or energy). Because of this historical treatment, low energy Compton scattering is occasionally referred to as Thomson scattering — even though the Thomson model itself is inconsistent (that is, elastic scattering implies coherency, but the Thomson model requires the electron to be free!)

Returning to the process in question (elastic nuclear scattering), we have the situation of a photon interacting with a nucleon in such a manner that a photon is re-emitted with the same energy. One sometimes refers to this as "Thomson scattering from a nucleus" in analogy to the low energy limit of Compton scattering.

B. Nuclear Resonance Scattering (II-C)

This effect is a type of inelastic nuclear scattering whereby the nucleus is raised to an excited level by absorbing a photon. The excited nucleus subsequently de-excites by emitting a photon of equal or lower energy.

C. Delbruck Scattering (III-B)

The phenomenon of the scattering of photons by the Coulomb field of a nucleus is called Delbruck scattering (also called nuclear potential scattering). It can be thought of as virtual pair production in the field of the nucleus — that is, pair production followed by annihilation of the created pair. The process is elastic.

D. Photomeson Production (IV-A)

Typical reactions:
$$\gamma + p \rightarrow \pi^+ + n$$
$$\gamma + p \rightarrow \pi^+ + \pi^- + p$$
$$\gamma + p \rightarrow \pi^o + \pi^+ + n$$
$$\gamma + n \rightarrow \pi^- + p$$

etc.

2.3 **Minor Processes**

A.   Rayleigh Scattering (I-B)

Rayleigh scattering (also called "electron resonance scattering") is an atomic process in which the incident photon is absorbed by a (tightly) bound electron. The electron is raised to a higher energy state, and a second photon of the same energy as the incident photon is then emitted, with the electron returning to its original state (this is not excitation, however). In effect, the recoil of the scattered photon is taken up by the atom as a whole with a very small energy transfer; so the photon loses negligible energy upon scattering. The process is elastic.

B.   Photonuclear Reactions (II-A)

Analogous to the photoelectric effect for electrons, a nucleus can absorb a photon and subsequently emit one or more nucleons — hence, the name "nuclear photoeffect." All such reactions have a threshold photon energy below which the reaction cannot occur. For the (γ, n) reaction, the cross section increases with increasing energy (above threshold), reaches a maximum value, and then decreases. This is referred to as the giant resonance, and is attributed to electric dipole absorption of the incident photon. In all cases, the maximum value of the total cross section for all photonuclear reactions is smaller than 5% of the total cross section of the same atom for Compton and pair-production interactions. This process is, therefore, not generally too important as a means of energy absorption. However, it can result in radioactive nuclei.

C.   Pair Production in the Field of an Electron (III-A-b)

This process is easier to understand after discussing pair production in the field of a nucleus. Thus, even though it is a minor effect, it will be discussed later.

2.4 Major Processes

A. Photoelectric Effect (I-A)

In the atomic photoeffect, a photon disappears and an electron is ejected from an atom. One should not visualize this interaction as occurring between a photon and an electron, but rather between a photon and an atom. In fact, a complete absorption type interaction cannot occur between a photon and a free electron since linear momentum will not be conserved.

Proof:

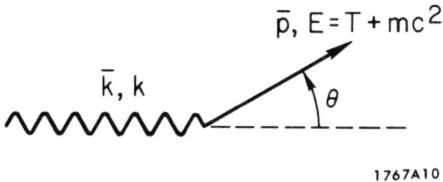

$\bar{p}$ = momentum of electron

mc^2 = rest mass of electron = 0.511 MeV

T = kinetic energy of electron

E = total energy of electron

$\bar{k}$ = momentum of photon ($k = |\bar{k}| = h/\lambda = h\nu/c$, so that if we work in $c = 1$ units, $k = h\nu > 0$)

In $c = 1$ units, the energy and momentum of a photon have the same magnitude. Hence,

$$\text{Conservation of Momentum: } \bar{k} = \bar{p}$$

$$\text{Conservation of Energy: } k + m = E.$$

Also

$$E^2 = p^2 + m^2 \text{ (invariance law)}.$$

Hence
$$(k+m)^2 = k^2 + m^2$$
$$= k^2 + m^2 + 2mk$$

This implies $2mk = 0$, hence either $m = 0$ or $k = 0$, which contradicts the assumptions that $m = mc^2 = 0.511$ MeV and $k > 0$. Thus, linear momentum is not conserved.

Even though the nucleus must absorb the momentum, it acquires very little kinetic energy due to its large mass.

Now clearly, the photoelectric effect can occur only if the incoming photon has an energy higher than the binding energy of the electron to be removed. We thus have a series of jumps in the curve of the absorption coefficient (or cross section), corresponding to the binding energy of the different shells. These energies are given approximately by Moseley's law:

$$E = 13.6 \frac{(Z-\sigma)^2}{n^2} \text{ (eV)} \qquad (2.4)$$

where Z = atomic number

σ = screening constant

n = quantum number of orbit such that $n = 1 \longrightarrow$ K series

$n = 2 \longrightarrow$ L series, etc.

Note that Moseley's law is essentially the energy of a Bohr orbit, modified by a screening constant.

The screening constant is approximately 3 for the K-shell and 5 for the L-shell. As an example, we can use Moseley's law to calculate the K and L absorption edges of lead ($Z = 82$), to get:

K-edge ($n = 1$) $\longrightarrow E = 13.6 \frac{(82-3)^2}{1^2}$ eV = 85 KeV

L-edge ($n = 2$) $\longrightarrow E = 13.6 \frac{(82-5)^2}{2^2}$ eV = 20 KeV

Whereas, the actual values are:

$$\text{K-edge: } 88.005 \text{ KeV}$$

$$\text{L-edge: } \begin{cases} L_1\text{-edge: } 15.855 \text{ KeV} \\ L_2\text{-edge: } 15.205 \text{ KeV} \\ L_3\text{-edge: } 13.041 \text{ KeV} \end{cases}$$

We see that the L-edge actually consists of three different numbers, as required by the quantum numbers

$$n = 2, \; l = 1, \; j = 3/2 \quad \text{(P-state)}$$
$$j = 1/2$$
$$l = 0, \; j = 1/2 \quad \text{(S-state)}$$

Because a third body (the nucleus) is required for momentum conservation, it makes sense that photoelectric absorption should increase rapidly with the binding energy of the electron. That is, the probability of this interaction is highest for those electrons most tightly bound. About 80% of the interactions involve the K-shell electrons. The order of magnitude of the photoelectric atomic-absorption coefficient is

$$\tau_{pe} \begin{cases} \sim Z^4/(h\nu)^3 & \text{low energy} \\ \sim Z^5/h\nu & \text{high energy} \end{cases}$$

That is, the photoelectric cross section decreases with increasing photon energy much more slowly at high photon energies.

The vacancy created by the ejection of an electron from the inner shells is filled by outer electrons falling into it (de-excitation) and this process may be accompanied by

a. emission of fluorescent radiation, or

b. Auger electron emission

c. or both.

The competition between the emission of a K x-ray and the emission of an Auger electron is described by the K fluorescence yield, which is defined as the number of K x-ray quanta emitted per vacancy in the K shell. The probability that a K x-ray will be emitted is nearly unity in high-Z elements and nearly zero in low-Z elements.[4]

Now, this brings up an interesting question of whether or not the Auger process should be considered as a process whereby a virtual fluorescent x-ray "converts," by means of a photoelectric interaction, before it escapes the atom. Clearly, the Auger process, from the discussion above, decreases in importance as Z increases. But, the photoelectric process increases with Z^4 (to Z^5)! Thus, it appears improbable that this is what happens. In addition, the nuclear analog to the Auger process—called "internal conversion" — provides evidence to support the conclusion that the conversion electron (or the Auger electron) is not due to an "internal photoelectric effect." It is observed experimentally that the $0 \rightarrow 0$ transition proceeds readily enough by internal conversion within the nuclear volume, although the emission of photons by the nucleus, in a $0 \rightarrow 0$ transition, is completely forbidden according to quantum mechanics.

B. Pair Production (III-A)

Pair production is the mechanism by which a photon is transformed into an electron-positron pair, also known as "materialization." The principle of conservation of momentum and energy prevents this from occurring in free space. There must be a nucleus or an electron present for this process to happen. In the center-of-mass system, the threshold for the materialization process is obviously $2mc^2 = 1.022$ MeV.

For the reaction $M_1 + M_2 \rightarrow M_3 + M_4 + M_5 + Q$, it can be shown from conservation of energy and momentum that the threshold energy for the reaction in the laboratory system is

$$T_{th}^{lab} = \frac{Q}{2M_2}[Q - 2(M_1 + M_2)]$$

when M_2 is at rest. In the pair production interaction $(\gamma + M \rightarrow M + m + m + Q)$,

$$M_1 = 0$$
$$M_2 = M_3 = M$$
$$M_4 = M_5 = m$$

so that

$$Q = -2m \left(= -T_{th}^{CM} \right)$$

and

$$T_{th}^{lab} = \frac{2m(m + M)}{M}$$

Thus

a. Pair production in the field of a nucleus of mass M:

$$M \gg m$$

$$T_{th}^{lab} \simeq \frac{2m}{M}(M) = 2m = 1.022 \text{ MeV}$$

b. Pair production in the field of an electron:

$$M = m$$

$$T_{th}^{lab} = \frac{2m}{m}(m + m) = 4m = 2.044 \text{ MeV}$$

1. **In the field of a nucleus (III-A-a)**

The presence of the nucleus guarantees conservation of momentum with negligible energy transfer to the nucleus. The atomic cross section for pair production in the neighborhood of a nucleus is proportional to Z^2. However, for photon energies above 20 MeV, one must use an "effective" Z in order to account for the screening of the true charge by atomic electrons. For low photon energies,

$$\kappa_n \sim \ln(h\nu)$$

For high energies,[5]

$$\kappa_n \sim \frac{7}{9X_0} \qquad (2.5)$$

where X_0 is the <u>radiation length</u> of the material (the definition of radiation length comes about in a natural way in describing the energy loss by an electron due to radiative (bremsstrahlung) processes — we will discuss this in Chapter 3).

The high energy approximation is quite useful for those people who work around a high energy electron accelerator ($h\nu \geq 100$ MeV). Generally, these people know the values for the radiation lengths of various materials, but do not have the absorption coefficients readily available. Just how good the approximation is, is shown in Table 2.2: (κ_n for 1000 MeV)

TABLE 2.2

Material	$X_0 (\text{g-cm}^{-2})$	$\frac{7}{9X_0}(\text{cm}^2\text{-g}^{-1})$	$\kappa_n(\text{cm}^2\text{-g}^{-1})$	% Difference
Pb	6.40	0.122	0.114	7
Cu	13.0	0.060	0.055	9
Fe	13.9	0.056	0.051	10
Al	24.3	0.032	0.028	14
C	43.3	0.018	0.014	29
H_2O	36.4	0.021	0.020	5

The fact that a quantity, X_0, that is defined in terms of a radiative process, can be used to evaluate a quantity associated with pair production, namely, κ_n, is not coincidental. If one writes the Feynman diagrams[9] for the two processes:

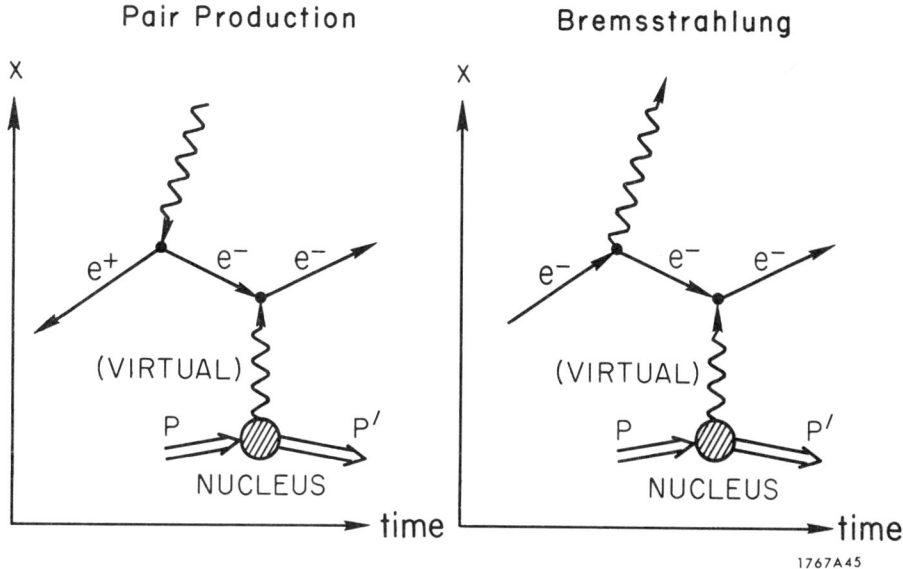

it becomes apparent that the two processes are identical — under the usual rule of changing the direction of the arrowhead and also changing the particle to its anti-particle. In other words, the derivation of the pair production and bremsstrahlung cross sections are essentially the same.

2. **In the field of an electron (III-A-b)**

When the recoil is absorbed by an electron, the threshold energy in the laboratory system is $4mc^2 = 2.044$ MeV, and there are two electrons and a positron acquiring appreciable momentum. In this case, the recoiling particle (electron) has considerable energy, so that the process is generally referred to as "triplet" production. At high photon energies the cross section for triplet production is about $1/Z$ times that for ordinary pair production. Thus, triplet production is of no consequence (relative to pair production) except for low-atomic-number materials.

Examples of both pair and triplet production are shown in the photographs*
(Fig. 2.1 and 2.2). Notice also the Compton interaction. The curvature of the
Compton electrons (due to the magnetic field that is being applied) helps identify
the positrons from the electrons.

In some applications, one must be perfectly correct in calculating the energy
absorption from pair production interactions, and therefore must account for the
annihilation of the positron with an atomic electron. Annihilation radiation assumes
a role analogous to scattered radiation in the Compton case and to fluorescence radiation in the photoelectric case. In most dosimetry applications, however, annihilation
radiation can be neglected because either the Compton effect dominates (i.e., pair
production is relatively small), or the fraction of the total pair production absorption
coefficient contribution due to annihilation is quite small. That is,

$$\kappa\left(1 - \frac{2mc^2}{h\nu}\right) \rightarrow \kappa \qquad (h\nu \gg 2mc^2)$$

Finally, the "characteristic" angle between the direction of motion of the
photon and one (or the other) of the electrons (±) is given by

$$\theta = \frac{mc^2}{h\nu}$$

Similarly, for the bremsstrahlung process,

$$\theta = \frac{mc^2}{E}$$

where E is the energy of the electron.

C. Compton Scattering (I-C)

When an incident photon is scattered by a loosely bound (or virtually free) electron,
the phenomenon is called Compton scattering. As was indicated earlier, this process
is an inelastic one in that some of the initial kinetic energy of the photon is needed in
order to overcome the binding energy of the electron to the atom, and therefore does
not appear as kinetic energy of the products. However, the process is treated as an

---

*40-inch bubble chamber (Stanford Linear Accelerator Center).

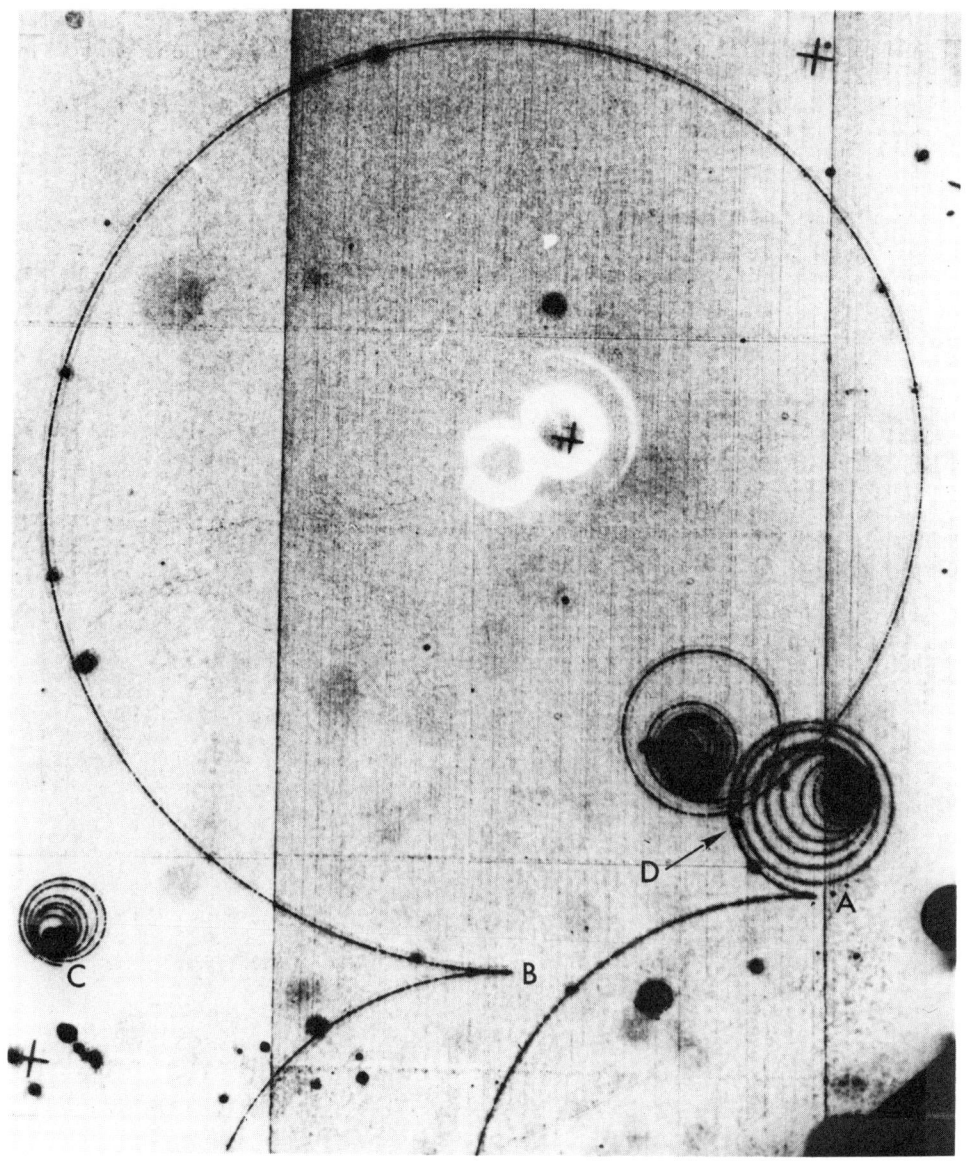

FIG. 2.1

Pair production in the field of a nucleus (A and B).
Compton interaction (C). Bremsstrahlung (D).

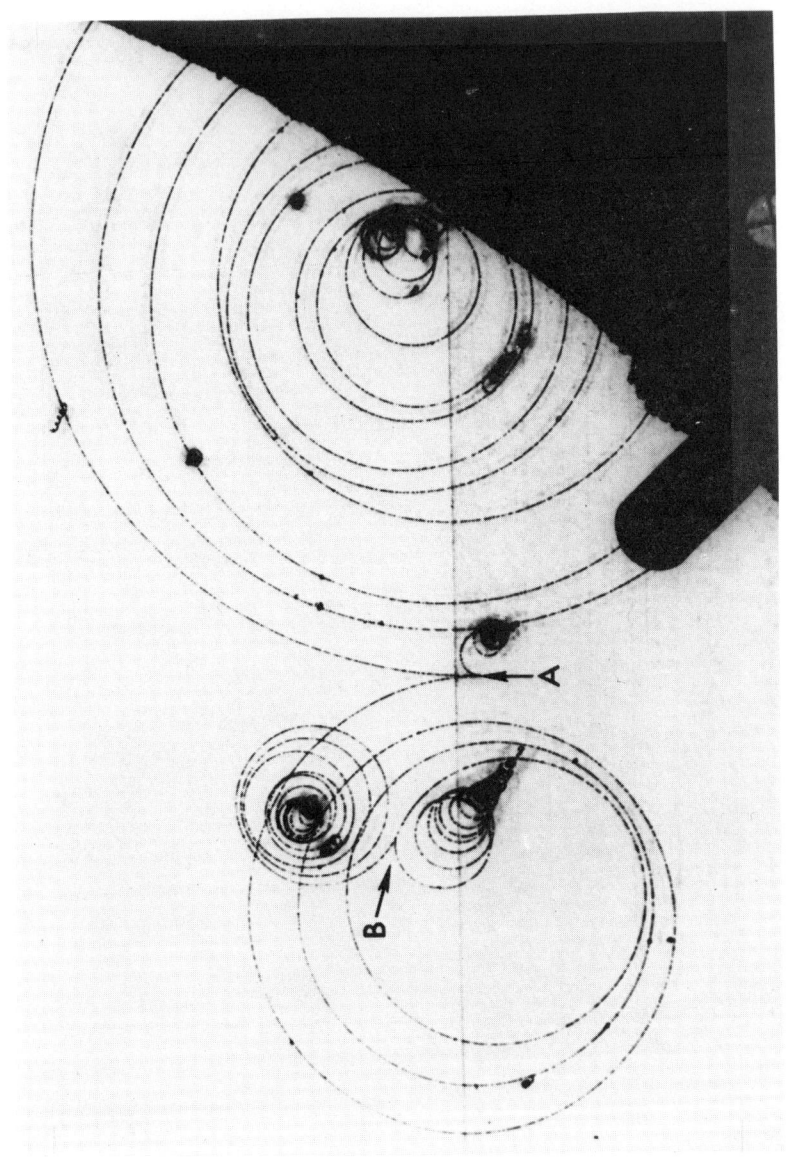

FIG. 2.2

An incident photon (no track) undergoes pair production in the field of an electron (triplet) at point A. The positron subsequently transfers a large amount of energy to an electron at point B. This type of interaction will be discussed in Chapter 3.

elastic one because this binding energy is small compared with the photon energy incident. This is a first order approximation and appropriate corrections are sometimes necessary for low energy photons or high Z materials (FBM, page 190).

The Compton process is described by the following diagram (in c = 1 units).

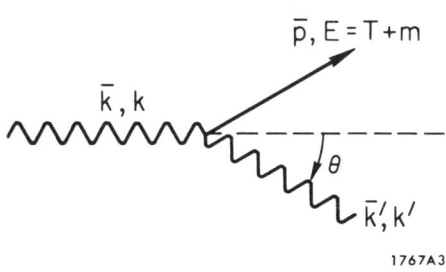

Conservation of momentum:

$$\bar{k} = \bar{k}' + \bar{p}$$

Conservation of energy:

$$k + m = E + k'$$

Invariance:

$$E^2 = p^2 + m^2$$

Hence,

$$E^2 = (\bar{k} - \bar{k}') \cdot (\bar{k} - \bar{k}') + m^2 \quad \text{(from C. of P.)}$$
$$= k^2 + k'^2 - 2kk' \cos\theta + m^2$$
$$= (k + m - k')^2 \quad \text{(from C. of E.)}$$

so that $\quad m(k - k') = kk'(1 - \cos\theta)$

or

$$\frac{1}{k'} - \frac{1}{k} = \frac{1}{m}(1 - \cos\theta) \qquad (2.6)$$

Now, since the right-hand side of this equation has units of reciprocal mass energy, we can go back to the "usual" notation by letting

$$m \rightarrow mc^2$$
$$k \rightarrow h\nu$$
$$k' \rightarrow h\nu'$$

which leads to the well known result

$$\lambda' - \lambda = \frac{h}{mc} (1 - \cos \theta) \qquad (2.7)$$

or alternately,

$$h\nu - h\nu' = h\nu \; \frac{\alpha(1 - \cos \theta)}{1 + \alpha(1 - \cos \theta)} = T \qquad (2.8)$$

where $\alpha = h\nu/mc^2$.

It is of great practical importance to note that the Compton shift in wavelength, in any particular direction, is independent of $h\nu$; whereas, the shift in energy is very dependent on $h\nu$. That is, high energy photons suffer a large energy change, but low energy photons do not. For $\theta = 90°$,

$$h\nu' = \frac{mc^2}{1 + mc^2/h\nu}$$

so that $h\nu'$ becomes a maximum when $h\nu \rightarrow \infty$, and therefore

$$h\nu' \leq 0.511 \text{ MeV}.$$

The total differential probability, $d\sigma/d\Omega$, for a photon to make a Compton collision such that the scattered photon is within a solid angle about theta, is given by the Klein-Nishina formula (ART, p. 102). Integrating over all angles leads to the total Compton cross section used in the mass attenuation coefficient, according to

$$\sigma = Z \int_{4\pi} \frac{d\sigma}{d\Omega} \, d\Omega \qquad \text{(barns/atom)}$$

where $d\sigma/d\Omega$ is in barns/electron - sr and Z is the atomic number.

The absorption component of the total differential cross section is obtained by weighting the total differential cross section by the fraction of energy carried off by the electron. That is,

$$\frac{d\sigma_a}{d\Omega} = \frac{d\sigma}{d\Omega} \frac{E(\theta)}{h\nu} .$$

The total Compton absorption coefficient can be obtained by integration over all solid angles as follows:

$$\sigma_a = \frac{1}{h\nu} \int_{4\pi} \frac{d\sigma}{d\Omega} E \, d\Omega$$

$$= \frac{1}{h\nu} \int_{4\pi} \frac{d\sigma}{d\Omega} d\Omega \frac{\int_{4\pi} \frac{d\sigma}{d\Omega} E \, d\Omega}{\int_{4\pi} \frac{d\sigma}{d\Omega} d\Omega} = \frac{1}{h\nu} \sigma \overline{E}$$

or

$$\sigma_a = \sigma \frac{\overline{E}}{h\nu}$$

Similarly, one can determine the scattering component. When integrated over all angles, we can obtain the result:

$$\sigma = \sigma_a + \sigma_s$$

2.5 Attenuation and Absorption

For use in calculating photon attenuation and absorption several macroscopic quantities have been developed from the cross sections for the processes discussed in this chapter. The ICRU has given official sanction to three coefficients (see Chapter 1):

Mass attenuation coefficient

$$\mu/\rho = \frac{1}{\rho}(\tau + \sigma + \sigma_R + \kappa) \tag{2.9}$$

Mass energy transfer coefficient

$$\mu_k/\rho = \frac{1}{\rho}\left[\tau(1-f) + \sigma\frac{\overline{E}}{h\nu} + \kappa\left(1 - \frac{2mc^2}{h\nu}\right)\right] \quad (2.10)$$

Mass energy absorption coefficient

$$\mu_{en}/\rho = \mu_k(1-G)/\rho \quad (2.11)$$

The units of these coefficients are cm^2/g and the symbols are the following:

τ = photoelectric cross section

σ = total Compton cross section

σ_R = Rayleigh cross section

κ = pair production cross section

f = fluorescent x-ray fraction

G = fraction of energy lost by secondary electrons in bremsstrahlung processes.

These coefficients will be referred to and used in subsequent chapters.

Two other coefficients often found in the literature are both called mass absorption coefficients and are approximations to the mass energy absorption coefficient:

$$\mu_a/\rho = \frac{1}{\rho}\left(\tau + \sigma\frac{\overline{E}}{h\nu} + \kappa\right) \quad (2.12)$$

$$\mu_{absn}/\rho = \frac{1}{\rho}\left[\tau + \sigma\frac{\overline{E}}{h\nu} + \kappa\left(1 - \frac{2mc^2}{h\nu}\right)\right] \quad (2.13)$$

These coefficients will not be used in this monograph. Tabulations of the various coefficients can be found in the literature.[6,7,8]

REFERENCES

1. U. Fano, L. V. Spencer, and M. J. Berger, Encylcopedia of Physics, Vol. XXXVIII/2, S. Flugge (ed.), (Berlin/Gottingen/Heidelberg, Springer, 1959).

2. Engineering Compendium on Radiation Shielding, Vol. I, "Shielding fundamentals and methods" (Springer-Verlag, New York, 1968); p. 185.

3. J. J. Thomson, Conduction of Electricity Through Gases (Cambridge University Press, London and New York, 1933).

4. R. D. Evans, The Atomic Nucleus (McGraw-Hill, New York, 1955); p. 565.

5. B. Rossi, High Energy Particles (Prentice-Hall, Inc., Englewood Cliffs, New Jersey, 1952).

6. G. W. Grodstein, "X-ray attenuation coefficients from 10 keV to 100 MeV," NBS Circular 583 (April 1957).

7. E. Storm and H. I. Israel, "Photon cross sections from 1 keV to 100 MeV for elements Z = 1 to Z = 100," At. Data and Nucl. Data Tables 7, 565 (1970).

8. J. H. Hubbell, "Photon cross sections, attenuation coefficients, and energy absorption coefficients from 10 keV to 100 GeV," NSRDS-NBS-29 (1969); Rad. Res. 70, 58 (1977).

9. R. B. Leighton, Principles of Modern Physics (McGraw-Hill, New York, 1959); pp. 669-679.

MAIN REFERENCES

(ART) F. H. Attix, W. C. Roesch, and E. Tochilin (eds.), Radiation Dosimetry, Second Edition, Volume I, Fundamentals (Academic Press, New York, 1968).

(FBM) J. J. Fitzgerald, G. L. Brownell, and F. J. Mahoney, Mathematical Theory of Radiation Dosimetry (Gordon and Breach, New York, 1967).

CHAPTER 3

CHARGED PARTICLE INTERACTIONS

3.1 Introduction

In the previous chapter we saw that photon interactions in matter resulted in the transfer of significant amounts of kinetic energy to electrons. This chapter will consider in detail the interactions of charged particles and particularly electrons as they move through a medium. Charged particles moving through a medium interact with the medium basically in three different ways: (1) by collision with an atom as a whole, (2) by collision with an electron, and (3) by radiative processes (bremsstrahlung). The mode of interaction is largely determined by the energy of the particle and the distance of closest approach of the particle to the atom with which it interacts.

A. If the distance of closest approach is large compared with atomic dimensions, the atom as a whole reacts to the field of the passing particle. The result is an excitation or ionization of the atom. The coulomb force is the major interaction force and the passing particle is considered a point charge. These distant encounters are also called soft collisions.

B. If the distance of closest approach is of the order of atomic dimensions, the interaction is between the moving charged particle and one of the atomic electrons. This process results in the ejection of an electron from the atom with considerable energy and is often described as a knock-on process, or hard collision. In general, the energy acquired by the secondary electron is large compared with the binding energy and the process can be treated as a free electron collision, but the intrinsic magnetic moment (spin) of the charged particle must be taken into account in the collision probability. Radiative processes can still be ignored but if the particles are identical, exchange phenomena occur and become especially important when the minimum distance of approach is of the order of the deBroglie wavelength, $\lambda = h/p$.

C. When the distance of closest approach becomes smaller than the atomic radius, the deflection of the particle trajectory in the electric field of the nucleus is the most important effect. This deflection process results in <u>radiative</u> energy losses and the emitted radiation (bremsstrahlung) covers the entire energy spectrum up to the maximum kinetic energy of the charged particle. But, quantum electrodynamics (QED) demands that

1. if radiation is emitted, it usually consists of a number of low-energy (soft) quanta such that

$$\sum_i (h\nu)_i \ll T \text{ (total KE of particle), and}$$

2. once in a while a photon may be emitted with energy comparable to the incident-particle energy.

3.2 Kinematics of the Collision Process*

We will discuss the collision process in an intermediate energy region where the interaction can be treated as a collision with a free electron.

Consider an elastic collision between a moving particle of mass M, total energy $E = T + M$ and momentum $\bar{p}$, and an electron at rest with mass m. The interaction

---

*The discussion of the collision kinematics and all subsequent probability formulas will be in $c = 1$ units. Thus, to return to cgs units replace m or M with mc^2 or Mc^2, respectively, wherever they appear.

can be described by the following figure

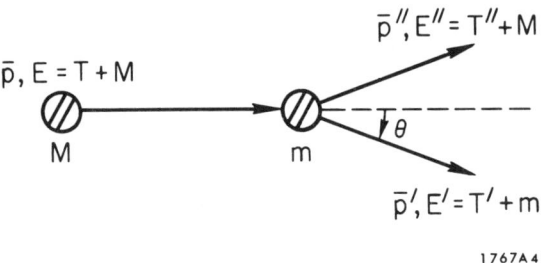

Conservation of Energy:
$$E + m = E' + E''$$

Conservation of Momentum:
$$\bar{p} = \bar{p}' + \bar{p}''$$

Invariance:
$$E''^2 = p''^2 + M^2$$

which lead to
$$E' = \frac{m[(E+m)^2 + p^2 \cos^2\theta]}{(E+m)^2 - p^2 \cos^2\theta} = T' + m$$

Hence,
$$T' = 2m \frac{p^2 \cos^2\theta}{[m + (p^2 + M^2)^{1/2}]^2 - p^2 \cos^2\theta} \quad (3.1)$$

$$= \text{K.E. of recoil electron.}$$

Now, T' is a maximum when $\theta = 0$, so that

$$T'_{max} = 2m \frac{p^2}{m^2 + M^2 + 2m(p^2 + M^2)^{1/2}} \quad (3.2)$$

This formula is identical to Eq. (2) of Barkas and Berger.[1]

For mesons and protons, $M \gg m$ so that two cases are of interest:

1. <u>High Energy Case</u>:

   For $\quad p \gg M^2/m$

   we have $\quad T'_{max} \simeq T$.

That is, a high energy meson or proton can be practically stopped by a head-on collision with a free electron.

2. Low Energy Case:

For
$$p \ll M^2/m$$

we have
$$T'_{max} \simeq 2m(p/M)^2 = 2m \frac{\beta^2}{1-\beta^2} = 2m\eta^2$$

where
$$\eta^2 \equiv \frac{\beta^2}{1-\beta^2} .$$

That is, the maximum energy transfer for a low energy meson or proton depends only on the particle velocity.

Barkas and Berger[1] point out that if the particle momentum is so great that the approximation $T'_{max} \simeq 2m\eta^2$ fails, the moving particle also probably cannot be treated as a point-charge. This implies that form-factor effects will then have to be included. It should be noted that even for the muon (the particle closest in mass to the electron) $M^2/m \simeq 20,000$ MeV. Consequently, for most attainable energies the low energy approximation will hold.

Now for the case of the electron, $M = m$, so that:

$$T'_{max} = 2m \frac{p^2}{m^2 + m^2 + 2m(p^2 + m^2)^{1/2}}$$

$$= \frac{p^2}{m + (p^2 + m^2)^{1/2}} \qquad (3.3)$$

But,
$$E^2 = (T + m)^2 = p^2 + m^2,$$

so that
$$T'_{max} = \frac{T(T + 2m)}{T + 2m} = T .$$

37

But since the two electrons are indistinguishable after the collision, by convention the one with the highest energy is considered the primary electron and so

$$T'_{max} \equiv T/2.$$

3.3 Collision Probabilities with Free Electrons (Knock-on Cross Sections)

The differential collision probability $\Phi_{col}(T, T')dT'dx$ is defined as the probability for a charged particle of kinetic energy T, traversing a thickness $dx(g\text{-}cm^{-2})$, to transfer an energy dT' about T' to an atomic electron (assumed free).

Note: In the notation of FBM,

$$\Phi_{col} dT' = \frac{N_0 Z}{A} d\sigma_H \quad (cm^2\text{-}g^{-1})$$

where the H refers to "hard" collisions.

A. Incident Electrons (Møller Cross Section)

For $T \gg m$ (c=1)

$$\Phi_{col}(T, T')dT' = 2Cm \frac{T^2 dT'}{(T-T')^2 (T')^2} \left[1 - \frac{T'}{T} + \left(\frac{T'}{T}\right)^2 \right]^2 \qquad (3.4)$$

= probability that either electron is in dT' about T'

where $\quad C = \pi N_0 (Z/A) r_0^2 = 0.150 \ (Z/A) \ (cm^2\text{-}g^{-1})$

A, Z = atomic weight, number

N_0 = Avogadro's number = 6×10^{23} atoms/mole

$r_0 = e^2/m = 2.82 \times 10^{-13}$ cm = classical radius of electron

Remark: One cannot distinguish between the primary and secondary electron. Therefore, Φ_{col} must be interpreted as leaving one electron at T' and the other at $T - T'$. All possible cases are accounted for with $0 \le T' \le T/2$, so that for electron-electron interactions, $T'_{max} = T/2$. Note that Φ_{col} is symmetric in both T' and $T - T'$. Figure 3.1 shows an electron interaction in which T' is approximately $T/2$.

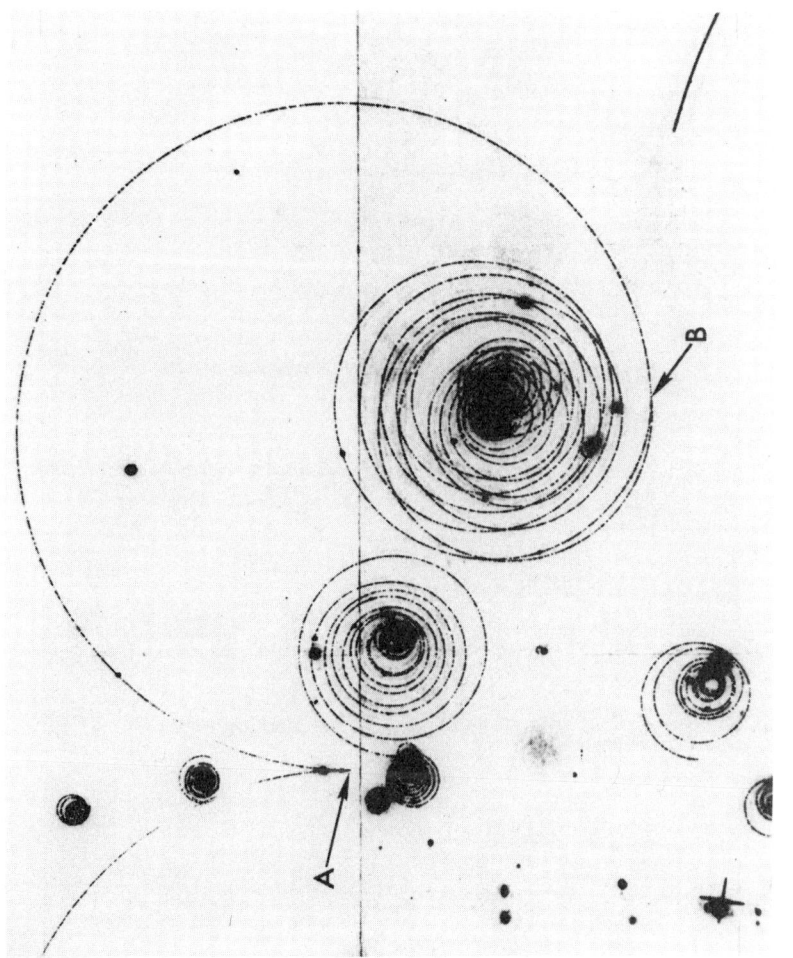

FIG 3.1

An incident photon (no track) undergoes a pair production interaction at point A. Subsequently the electron (identified by its curvature in relation to the Compton electrons) transfers approximately half its energy to a second electron at point B.

B. Incident Positrons (Bhabha Cross Sections)

For $T \gg m$

$$\Phi'_{col}(T,T')dT' = 2Cm\frac{dT'}{(T')^2}\left[1 - \frac{T'}{T} + \left(\frac{T'}{T}\right)^2\right]^2 \tag{3.5}$$

= probability that the electron is in dT' about T'

and

$$\Phi''_{col}(T,T')dT' = 2Cm\frac{dT'}{(T-T')^2}\left[1 - \frac{T'}{T} + \left(\frac{T'}{T}\right)^2\right]^2 \tag{3.6}$$

= probability that the positron is in dT' about T'

so that

$$\Phi_{col}(T,T')dT' = \left[\Phi'_{col}(T,T') + \Phi''_{col}(T,T')\right]dT' \tag{3.7}$$

= probability that either the positron or the electron

is in dT' about T'.

C. Heavy Incident Particles of Spin One-Half (e.g., Protons and Muons) (Bhabha, Massey and Corben Cross Section)

For $T \gg m$

$$\Phi_{col}(T,T')dT' = \frac{2Cm}{\beta^2}\frac{dT'}{(T')^2}\left[1 - \beta^2\frac{T'}{T'_{max}} + \frac{1}{2}\left(\frac{T'}{T+M}\right)^2\right] \tag{3.8}$$

D. Heavy Incident Particles of Spin Zero (e.g., Alpha Particles and Pions) (Bhabha Cross Section)

For $T \gg m$

$$\Phi_{col}(T,T')dT' = \frac{2Cm}{\beta^2}\frac{dT'}{(T')^2}\left[1 - \beta^2\frac{T'}{T'_{max}}\right] \tag{3.9}$$

(Note: for alpha particles one must multiply by $z^2 = 4$, since all formulas above assume $z = 1$).

E. Rutherford Formula

When $T' \ll T'_{max}$ (i.e., distant collisions with little energy transfer). The above formulas (3.4, 3.7, 3.8, 3.9) reduce to

$$\Phi_{col}(T,T')dT' = \frac{2Cm}{\beta^2} \frac{dT'}{(T')^2} \tag{3.10}$$

which is known as the Rutherford formula (not to be confused with the Rutherford scattering formula for the same process -- the elastic scattering of charged particles).

The above expression gives the collision probability for all particles and depends only on the energy of the secondary electron, T', and on the velocity of the primary particle. It can be derived rather easily using classical mechanics.

Consider a charged particle moving past a free electron as indicated below:

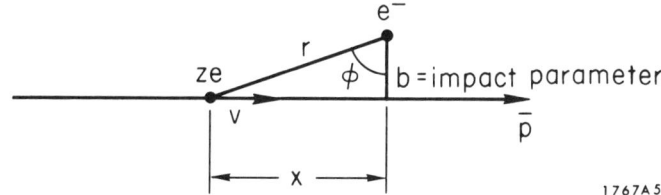

The momentum transferred to the electron, $\bar{p}'$, is calculated from

$$\bar{p}' = \int \bar{F}\, dt \quad \text{(time integration over the force)}$$

We are only interested in the perpendicular force, since the parallel forces cancel, so that

$$F = F_\perp = \frac{ze^2}{r^2} \cos\phi$$

$$F = \frac{ze^2 b}{(x^2 + b^2)^{3/2}}$$

Now,

$$x = vt$$

so that

$$dt = \frac{1}{v} dx$$

and therefore

$$|\bar{p}'| = \int_{-\infty}^{\infty} \frac{ze^2 b}{(x^2+b^2)^{3/2}} \frac{dx}{v} = \frac{2ze^2}{bv}$$

The energy transferred to the electron is

$$T' = \frac{|\bar{p}'|^2}{2m} = \frac{2z^2 e^2}{mb^2 v^2}$$

or

$$b^2 = \frac{2e^4}{m\beta^2 T'}$$

so that

$$|2bdb| = \frac{2e^4}{m\beta^2 (T')^2} dT'$$

for a $z = 1$ charge (incident particle). Now, the probability of a collision with impact parameter in db about b in a thickness dx is given by

$$F(b) db\,dx = 2\pi\, bdb \frac{N_0 Z}{A} dx = \Phi_{col} dT' dx$$

or

$$\Phi_{col}(T, T') dT' = \frac{2\pi e^4 N_0 Z}{m\beta^2 (T')^2 A} dT'$$

But,

$$r_0 = e^2/m$$

and

$$C = \pi N_0 \frac{Z}{A} r_0^2$$

so that

$$\Phi_{col}(T, T') dT' = \frac{2Cm}{\beta^2} \frac{dT'}{(T')^2} \quad (cm^2 - g^{-1})$$

The derivation of Rutherford's formula presented above brings out the physical basis for the dependence of Φ_{col} on the various factors in the formula:

1. The factor C expresses the proportionality of the collision probability to the electron density.

2. The factor $1/\beta^2$ expresses the dependence of the energy transfer on the collision time.

3. The factor $1/(T')^2$ expresses the fact that collisions with large impact parameters are more likely than collisions with small impact parameters.

3.4 Ionization Loss (Energy Loss by Collision)

So far we have restricted the discussion to collision probabilities of charged particles via hard collisions. In the total picture of charge-particle collisions, hard collisions are comparatively rare and do not have much influence upon the most probable energy loss. However, this should not be interpreted to mean that they are unimportant, since each hard collision carries away a relatively large amount of energy when it does occur.

The <u>average energy loss</u> per unit path length (also known as the <u>average stopping power</u>) from ionization (and excitation) is given by

$$\left.\frac{dT}{dx}\right)_{col} = \left.\frac{dT_S}{dx}\right)_{col} + \left.\frac{dT_H}{dx}\right)_{col}$$

where H means "hard" (close) and S means "soft" (distant). This can be written

$$\left.\frac{dT}{dx}\right)_{col} = \int_{T'_{min}}^{H} T' \phi_{col}^S \, dT' + \int_{H}^{T'_{max}} T' \phi_{col}^H \, dT' \quad (\text{MeV-cm}^2\text{-g}^{-1})$$

where

$\phi_{col}^H = \phi_{col}$ given in the formulas in 3.3.

ϕ_{col}^S = collision cross section for soft-collisions (not derived here).

H = energy transfer above which collisions can be considered hard.

Although not absolutely correct, let us now make the assumption that

$$\phi_{col}^S = \phi_{col}^H = \text{Rutherford formula (Eq. (3.10))}$$

$$\left(\frac{dT}{dx}\right)_{col} = \int_{T'_{min}}^{T'_{max}} T' \phi_{col}^{Ruth}(T, T')dT' = \frac{2Cm}{\beta^2} \int_{T'_{min}}^{T'_{max}} dT'/T' = \frac{2Cm}{\beta^2} \ln\left(\frac{T'_{max}}{T'_{min}}\right)$$

Now, it can be shown from quantum-mechanics that $T'_{max}/T'_{min} = (2mv^2/I)^2$ where I is the mean excitation energy. Thus,

$$\left(\frac{dT}{dx}\right)_{col} = \frac{4Cm}{\beta^2} \ln\left(\frac{2m\beta^2}{I}\right) \quad \text{in units of } c = 1.$$

Although not correct, it does indicate the general features of the theory. (Note: Again, this expression holds for $z = 1$ particle. For particles with charge z, multiply above (and future) stopping power formulas by z^2).

Now, the soft-collision stopping power, as derived by Bethe,[2] is

$$\left(\frac{dT_S}{dx}\right)_{col} = \frac{2Cm}{\beta^2} \left\{\ln\left(\frac{2m\beta^2 H}{I^2(1-\beta^2)}\right) - \beta^2\right\}. \qquad (3.11)$$

The derivation of (3.11) will not be presented here because of the difficulty that comes about because of the binding of the electrons to the atom. This shows up in the stopping power formula as the quantity I. Equation (3.11) applies for electrons as well as heavy charged particles.

We can calculate quite easily the hard-collision term for the case of a heavy (spin zero) particle. That is, (from Eq. (3.9))

$$\left(\frac{dT_H}{dx}\right)_{col} = \int_H^{T'_{max}} T' \phi_{col}^H dT' = \frac{2Cm}{\beta^2} \int_H^{T'_{max}} \frac{dT'}{T'}\left(1 - \beta^2 \frac{T'}{T'_{max}}\right)$$

$$= \frac{2Cm}{\beta^2}\left\{\ln\left(\frac{T'_{max}}{H}\right) - \beta^2\left(1 - \frac{H}{T'_{max}}\right)\right\}$$

and for $H \ll T'_{max}$

$$\left(\frac{dT_H}{dx}\right)_{col} = \frac{2Cm}{\beta^2}\left\{\ln\left(\frac{T'_{max}}{H}\right) - \beta^2\right\}.$$

So that upon adding the soft and hard terms:

$$\left.\frac{dT}{dx}\right)_{col} = \frac{2Cm}{\beta^2}\left\{\ln\left(\frac{2m\beta^2 T'_{max}}{I^2(1-\beta^2)}\right) - 2\beta^2\right\} \text{ (MeV-cm}^2\text{-g}^{-1}) \quad (3.12)$$

This relation applies to heavy charged particles (M≫m) with energy and charge fulfilling the Born approximation condition

$$\frac{2Zz}{137} \ll \beta \ .$$

At this point, certain modifications must be made to the basic formula to correct for various atomic effects. The first of these effects is known as the <u>polarization</u> (or <u>density</u>) effect. Up to this point, we have considered the collision process as occurring between the charged particle and isolated atoms. This is valid to a great extent when the absorbing medium is a gas. When the electron travels in a condensed medium, the atoms can be considered isolated only in the case of close collisions. However, for distant collisions we must consider the electrical polarization of the medium in which the particle moves. The dielectric constant of the medium weakens the electric field acting at a distance from the atom, causing a decrease of the energy transfer to atoms located far from the particle, and hence a decrease in the mass stopping power (soft-collision term).

Thus, in case of a medium in two phases of different densities, such as water and vapor, the lower density phase has a higher mass stopping power and hence the name "density effect" — this effect is appreciable, however, only for relativistic velocities. The most extensive treatment of this is that of Sternheimer.[3]

Another important effect of the dielectric constant is the production of Cerenkov radiation. This effect accounts for part of the relativistic correction to energy loss by distant collisions. The density effect and the Cerenkov light are interrelated, both being functions of the dielectric constant of the medium, and hence, are generally treated together.[3]

A second smaller correction is necessary because the atomic electrons will contribute less to the stopping power if the particle velocity is comparable to the velocity of the electron in its orbit. This shell correction can be as much as 10% for low energy heavy charged particles but is less than 1% for electrons of energies greater than 0.1 MeV and is a maximum of ~10% at an electron energy of about 2 keV. Consequently, shell corrections are generally ignored for electron stopping powers.

Considering all of these corrections, the final stopping power formula for a singly-charged particle heavier than an electron is

$$\left(\frac{dT}{dx}\right)_{col} = \frac{2Cm}{\beta^2}\left\{\ln\left(\frac{2m\beta^2 T'_{max}}{I^2(1-\beta^2)}\right) - 2\beta^2 - \delta - U\right\} \text{ (MeV-cm}^2\text{-g}^{-1}) \qquad (3.13)$$

where

δ = density effect correction[3]

U = shell correction term[3]

Equation (3.13) is equivalent to Eq. (1) of Barkas and Berger.[1]

The overall picture, then, is as follows:

1. The initial behavior of the ionization loss, given by Eq. (3.13), is that it starts decreasing proportional to β^2.

2. The logarithmic term containing the factor $1/(1 - \beta^2)$ causes a slow increase in the relativistic region (as the maximum effective impact parameter increases). The point at which the slope of dT/dx changes is known as minimum ionization. It occurs approximately at $T_{min} \sim 3M$.

3. The increase tends to flatten out into a plateau as the polarization effects become increasingly more significant. This plateau is of the order of 2 MeV-cm^2-g^{-1}.

Finally, one can go through a similar analysis for incident electrons and positrons. In particular, the soft collision formula ϕ_{col}^S, as given by Bethe,[2] is still correct. One need only to use the proper hard collision formula to obtain:

$$\left.\frac{dT}{dx}\right)_{col} = \frac{2Cm}{\beta^2} \left\{ \ln\left[\frac{\tau^2(\tau+2)}{2(I/m)^2}\right] + F^{\pm}(\tau) - \delta \right\} \quad (\text{MeV-cm}^2\text{-g}^{-1}) \qquad (3.14)$$

where

$$F^-(\tau) = 1 - \beta^2 + [\tau^2/8 - (2\tau+1)\ln 2]/(\tau+1)^2 \qquad (3.15)$$

for electrons and

$$F^+(\tau) = 2\ln 2 - \frac{\beta^2}{12}\left[23 + \frac{14}{\tau+2} + \frac{10}{(\tau+2)^2} + \frac{4}{(\tau+2)^3}\right] \qquad (3.16)$$

for positrons and where

$$\tau \equiv T/m$$

$$\delta = \text{density effect correction}^3$$

Stopping power values using Eq. (3.14) have been published by Berger and Seltzer.[4]

3.5 Restricted Stopping Power (LET)

For some applications the energy deposited by a charged particle in a region of specified dimensions about its track is of interest. The basic stopping power formula is used but we must exclude the energy escaping from the region of interest in the form of fast knock-on electrons (delta rays). The expression for the restricted mean collision loss for electrons and positrons (LET_Δ) is:

$$L^{\pm}(\tau, \Delta) = \frac{2Cm}{\beta^2}\left\{\ln\left[\frac{2(\tau+2)}{(I/m)^2}\right] + F^{\pm}(\tau, \Delta) - \delta\right\} \qquad (3.17)$$

for electrons

$$F^-(\tau, \Delta) = -1 - \beta^2 + \ln[(\tau-\Delta)\Delta] + \tau/(\tau-\Delta)$$
$$+ [\Delta^2/2 + (2\tau+1)\ln(1-\Delta/\tau)]/(\tau+1)^2 \qquad (3.18)$$

and for positrons

$$F^+(\tau,\Delta) = \ln(\tau\Delta) - \frac{\beta^2}{\tau}\left[\tau + \Delta - \frac{5\Delta^2/4}{\tau+2} + \frac{(\tau+1)(\tau+3)\Delta + (\Delta^3/3)}{(\tau+2)^2}\right.$$
$$\left. - \frac{[(\tau+1)(\tau+3)+3]\Delta^2/4 - \Delta^3\tau/3 + \Delta^4/4}{(\tau+2)^3}\right] \qquad (3.19)$$

In this formulation Δ is the kinetic energy of the delta ray which just escapes the region of interest. For an electron of energy τ passing through matter the maximum energy transferred to delta rays is $\tau/2$. By inserting $\Delta = \tau/2$ in the above equation for $L^-(\tau,\Delta)$ it can easily be shown that

$$L^-(\tau,\tau/2) = \left.\frac{dT}{dx}\right)_{col} ,$$

which is also called LET_∞ (or unrestricted stopping power).

3.6 Compounds

Often one needs to know the stopping power of compounds rather than pure elements. Stopping power can be calculated to a first approximation using Braggs additivity rule:

$$\frac{dT}{dx} = \sum_j \epsilon_j \left.\frac{dT}{dx}\right._j$$

where ϵ_j is the weight fraction of element j.

Since the Bragg additivity rule does not take into account the change of the electronic configuration in going from an element to a compound some error will be involved in the calculation. These errors will normally be of the order of a few percent and will be most serious for low energies.

3.7 Gaussian Fluctuations in the Energy Loss by Collision

Particles of a given kind and of a given energy do not all lose exactly the same amount of energy in traversing a given thickness of material. The actual energy loss is a statistical phenomenon and fluctuates around the average value as calculated above. Only heavy charged particles will be considered here since high energy electrons lose energy substantially by radiative collisions.

Let $\omega(T_0, T, x)dT$ represent the probability that a particle of initial energy T_0 has an energy in dT about T after traversing a thickness of $x(g\text{-}cm^{-2})$ of matter. Rossi[5] gives the following equation for $\omega(T_0, T, x)$:

$$\omega(T_0, T, x+dx) - \omega(T_0, T, x) = -\omega(T_0, T, x) \int_0^\infty \Phi_{col}(T, T') dT'$$
$$+ dx \int_0^\infty \omega(T_0, T+T', x) \Phi_{col}(T+T', T') dT' \quad (3.20)$$

where

$$\Phi_{col}(T, T') = 0 \text{ for } T' > T'_{max} \text{ and } \omega(T_0, T, x) = 0 \text{ for } T > T_0.$$

With the following assumptions:

1. $k_{col} \equiv \left(\dfrac{dT}{dx}\right)_{col} = \int_0^T T' \Phi_{col}(T, T') dT' = \text{constant}$

2. $T_a = T_0 - xk_{col} = \text{average energy at } x$

3. $\Phi_{col}(T+T', T') = \Phi_{col}(T, T') = \Phi_{col}(T') \text{ only}$

4. $\omega(T_0, T+T', x)$ varies only slightly so that one can expand in a power series of T' about T, and neglect terms beyond second order.

One obtains

$$\frac{\partial \omega}{\partial x} = k_{col} \frac{\partial \omega}{\partial T} + \frac{1}{2} \rho^2 \frac{\partial^2 \omega}{\partial T^2} \quad (3.21)$$

where

$$\rho^2 \equiv \int_0^\infty (T')^2 \Phi_{col}(T, T') dT'$$

To solve this, we introduce the Fourier transform pair

$$\overline{\omega}(x, \alpha) = \frac{1}{\sqrt{2\pi}} \int_{-\infty}^\infty \omega(x, T) e^{-i\alpha T} dT$$

$$\omega(x, T) = \frac{1}{\sqrt{2\pi}} \int_{-\infty}^\infty \overline{\omega}(x, \alpha) e^{i\alpha T} d\alpha$$

where we have temporarily dropped the T_0 for convenience. The Fourier transform of Eq. (3.21) is:

$$\frac{d\bar{\omega}}{dx} = i\alpha\, k_{col}\, \bar{\omega} - \frac{1}{2}\rho^2 \alpha^2 \bar{\omega}$$

$$\therefore \bar{\omega}(x,\alpha) = \bar{\omega}(0,\alpha) \exp\left[\left(i\alpha\, k_{col} - \frac{1}{2}\rho^2\alpha^2\right)x\right]$$

Now,

$$\omega(0, T) = \delta(T_0 - T) \quad (\text{i.e., single incident particle of energy } T_0)$$

so that

$$\bar{\omega}(0,\alpha) = \frac{1}{\sqrt{2\pi}} \int_{-\infty}^{\infty} \delta(T_0 - T) e^{-i\alpha T}\, dT = \frac{1}{\sqrt{2\pi}} e^{-i\alpha T_0}$$

Therefore,

$$\bar{\omega}(x,\alpha) = \frac{1}{\sqrt{2\pi}} \exp\left[-\left(i\alpha T_a + \frac{1}{2}\rho^2\alpha^2 x\right)\right]$$

where

$$T_a = T_0 - x\, k_{col}$$

And,

$$\omega(x, T) = \frac{1}{\sqrt{2\pi}} \int_{-\infty}^{\infty} \bar{\omega}(x,\alpha)\, e^{i\alpha T}\, d\alpha$$

$$= \frac{1}{2\pi} \int_{-\infty}^{\infty} \exp\left[-\left(i\alpha T_a + \frac{1}{2}\rho^2\alpha^2 x\right)\right] e^{i\alpha T}\, d\alpha$$

$$= \frac{1}{2\pi} e^{-(T-T_a)^2/2\rho^2 x} \int_{-\infty}^{\infty} \exp\left[-\frac{1}{2}\rho^2 x\left[\alpha - \frac{i(T-T_a)}{\rho^2 x}\right]^2\right] d\alpha$$

where we have completed the square.

Now, this integral can be accomplished by choosing the rectangular contour

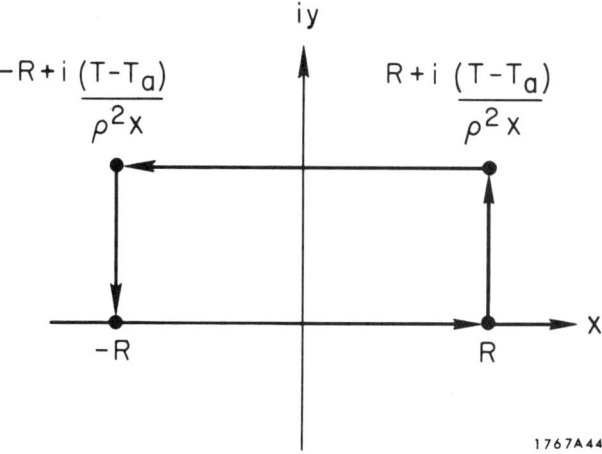

By Cauchy's theorem, the integral around this closed path is zero because the integrand is analytic at every point within and on C. As R becomes very large, the integrals along the vertical parts are seen to approach zero, and it follows that

$$\int_{-\infty}^{\infty} \exp\left[-\frac{1}{2}\rho^2 x\left[\alpha - \frac{i(T-T_a)}{\rho^2 x}\right]^2\right] d\alpha = \int_{-\infty+i(T-T_a)/\rho^2 x}^{\infty+i(T-T_a)/\rho^2 x} \exp\left[-\frac{1}{2}\rho^2 x\left[\alpha - \frac{i(T-T_a)}{\rho^2 x}\right]^2\right] d\alpha$$

$$= \sqrt{\frac{2}{\rho^2 x}} \int_{-\infty}^{\infty} e^{-u^2} du$$

$$= \sqrt{\frac{2\pi}{\rho^2 x}}$$

where

$$u = \sqrt{\frac{1}{2}\rho^2 x}\left[\alpha - \frac{i(T-T_a)}{\rho^2 x}\right]$$

Hence

$$\omega(T_0, T, x) = \frac{1}{\sqrt{2\pi \rho^2 x}} e^{-(T-T_a)^2/2\rho^2 x} \qquad (3.22)$$

Therefore, when all of the above conditions are fulfilled, the distribution function ω at the depth x is a Gaussian function of T with a maximum at T_a and having a half-width of

$$\sigma = \rho \sqrt{x}$$

The most probable energy is defined as the value of T for which the function $\omega(T_0, T, x)$ is a maximum. We see that this occurs at $T = T_a$, as expected.

Now, using the ϕ_{col}^H formula for spin zero particles, Eq. (3.9), (the other formulas could have been used as well), we have, *

$$\rho^2 = \int_0^\infty (T')^2 \, \phi_{col}^H(T, T') dT' = \frac{2Cm}{\beta^2} \int_0^{T'_{max}} \left(1 - \beta^2 \frac{T'}{T'_{max}}\right) dT' = \frac{2Cm T'_{max}}{\beta^2} \left(1 - \frac{\beta^2}{2}\right)$$

From experiment, the conditions for the validity of the Gaussian solution can be expressed by saying that

$$T'_{max} \ll \sigma \ll T_a \text{ (or } T_0 - T_a)$$

so that

$$\frac{2Cmx}{\beta^2 T'_{max}} \left(1 - \frac{\beta^2}{2}\right) = G\left(1 - \frac{\beta^2}{2}\right) \gg 1$$

In other words, we have a Gaussian distribution provided that

$$G = \frac{2Cmx}{\beta^2 T'_{max}}$$

is large.

---

*The expression for ρ^2 contains the factor $(T')^2$, whereas the expression for k_{col} contains the factor T'. Therefore, distant collisions are much less important in the computation of ρ^2 than they are in the computation of k_{col}, and we assume that $\phi_{col} \to \phi_{col}^H$ for all values of T' down to $T' = 0$.

For thin absorbers (i.e., small x) and/or high energies (so that T'_{max} is large), G is not a large quantity and one cannot consider the fluctuations as Gaussian.

3.8 Landau Fluctuations in the Energy Loss by Collision

When G is not large, one cannot replace the integro-differential Eq. (3.20) by the partial differential Eq. (3.21), and the determination of ω becomes a difficult mathematical task. Using Laplace transforms, Landau[6] has obtained a solution of the integro-differential equation that is valid when G is less than about 0.05. A complete solution has been given by Symon.[7] The most probable energy loss, ϵ_p, is obtained from the most probable energy, T_p, according to

$$\epsilon_p = T_0 - T_p = \frac{2Cmx}{\beta^2}\left[\ln\left(\frac{4Cm^2x}{(1-\beta^2)I^2}\right) - \beta^2 - \delta + j\right] \quad (3.23)$$

where j is a function of the parameter G and of the particle velocity β, and where δ is the density effect correction. For high energy particles traversing a thin absorber (i.e., $G \leq 0.05$)[23]

$$j \to 0.198$$

Now, since the probability of collision decreases with increasing energy transfer, that is,

$$\phi^H_{col} dT' \propto \frac{dT'}{(T')^2} ,$$

the energy-loss distribution is asymmetrical with a long tail on the high-energy side, corresponding to infrequent collisions with large energy transfer. This is called the Landau distribution.

3.9 Radiative Processes and Probabilities

The treatment of electron energy loss by radiative photon emission (bremsstrahlung) is influenced by the distance from the nucleus at which the radiative loss occurs. Radiative energy loss is caused by an acceleration (generally in the form of a change in direction) of the charged particle under the influence of the electric field of a nearby nucleus. If the distance of approach is large compared with the nuclear radius ($> 10^{-13}$ cm) but small compared with the atomic radius ($< 10^{-8}$ cm), the field can be considered that of a point charge Ze at the center of the nucleus. On the other hand, if the distance of approach is of the order of the atomic radius, or larger, the screening of the field of the nucleus by the atomic electrons must be considered. One might consider a third process whereby the distance of approach is of the order of the nuclear radius. As it turns out, in practice radiative processes take place at distances far from the nucleus so that we do not need to consider this.

According to the theory developed by Bethe and Heitler[8] (and summarized by Rossi[5]) based on the Fermi-Thomas atomic model the influence of screening on a radiative process depends on the recoil momentum of the atom in the process. The effect of screening on a radiative process in which an electron of initial total energy E(= T + m) produces a photon of energy $h\nu$ is measured by the quantity:

$$\gamma = 100 \; \frac{mh\nu}{E(E-h\nu)} \; Z^{-1/3} \tag{3.24}$$

It is seen that γ is an explicit function of the electron energy. When the energy E is small, γ is large and the screening may be neglected. When the electron energy is large, γ is small and the screening is nearly complete. Since the probability, $\phi_{rad}(T, h\nu) \; d(h\nu)dx$ for an electron of kinetic energy T to produce a photon in $d(h\nu)$ about $h\nu$ in traversing $dx(g\text{-}cm^{-2})$ is dependent on the screening effect, no single expression can be written for this probability. The radiation probability will be given here for two cases, no screening and complete screening with the restriction that E $\gg$ m.

No screening ($\gamma \gg 1$)

$$\phi_{rad}^n(T, h\nu)\, d(h\nu) = 4\alpha \frac{N_0}{A} Z^2 r_0^2 \frac{d(h\nu)}{h\nu} \left[1 + \left(\frac{E'}{E}\right)^2 - \frac{2}{3}\frac{E'}{E}\right]$$

$$\times \left[\ln\left(\frac{2EE'}{mh\nu}\right) - \frac{1}{2}\right] (cm^2\text{-}g^{-1}) \quad (3.25)$$

Complete screening ($\gamma \approx 0$)

$$\phi_{rad}^n(T, h\nu)\, d(h\nu) = 4\alpha \frac{N_0}{A} Z^2 r_0^2 \frac{d(h\nu)}{h\nu} \left\{\left[1 + \left(\frac{E'}{E}\right)^2 - \frac{2}{3}\frac{E'}{E}\right]\right.$$

$$\left.\times \left[\ln 183\, Z^{-1/3}\right] + \frac{1}{9}\frac{E'}{E}\right\} (cm^2\text{-}g^{-1}) \quad (3.26)$$

Note: $E' = E - h\nu$

$E = T + m \cong T$

α = fine structure constant = $1/137$

n = refers to "nucleus."

These probabilities are derived using the Born approximation which is valid only for elements where $Z/137 \ll 1$. For elements of high Z it can be shown that the Born approximation error is proportional to $(Z/137)^2$. The absolute error can be determined only by measurement. Experimentally it has been found that bremsstrahlung production from high Z materials is of the order of 5 to 10% higher than predicted by the theory.

Radiation energy loss by charged particles is also possible in the field of the atomic electrons (again, however, we only consider incident electrons). If the electron energy is such that screening may be neglected (and considering all of the electrons of the atom together), the probability of radiative energy loss is given by

$$\phi_{rad}^e = \frac{1}{Z} \phi_{rad}^n \; .$$

Therefore, the total probability is:

$$\Phi_{rad} d(h\nu) = \left[\Phi_{rad}^n + \Phi_{rad}^e\right] d(h\nu)$$

$$= 4\alpha \frac{N_0}{A} Z(Z+1) r_0^2 \frac{d(h\nu)}{h\nu} \left[1 + \left(\frac{E'}{E}\right)^2 - \frac{2}{3}\frac{E'}{E}\right]\left[\ln\frac{2EE'}{mh\nu} - \frac{1}{2}\right] (cm^2\text{-}g^{-1}) \quad (3.27)$$

For complete screening (and considering all of the electrons of the atom together),

$$\Phi_{rad}^e(T, h\nu) d(h\nu) = 4\alpha \frac{N_0}{A} Z r_0^2 \frac{d(h\nu)}{h\nu} \left[1 + \left(\frac{E'}{E}\right)^2 - \frac{2}{3}\frac{E'}{E}\right]\left[\ln 1440\, Z^{-2/3}\right]$$

$$+ \frac{1}{9}\frac{E'}{E} \quad (cm^2\text{-}g^{-1}) \quad (3.28)$$

Neglecting the 1/9 (E'/E) term the ratio of $\Phi_{rad}^e/\Phi_{rad}^n$ is proportional to 1/Z. The following table gives some comparisions:

Table 3.1

Z	1	10	92
$\Phi_{rad}^e/\Phi_{rad}^n$	1.40	0.129	0.0122
η(MeV) nuclei	87.	40.	19.3
η(MeV) electrons	490.	105.	24.

(η = energy required to obtain 90% of asymptotic value of Φ_{rad}.)

It is obvious that radiation energy losses in the field of electrons are important only for very high energy electrons in low Z materials. We can therefore write

$$\Phi_{rad} d(h\nu) = \left[\Phi_{rad}^n + \Phi_{rad}^e\right] d(h\nu)$$

$$= 4\alpha \frac{N_0}{A} Z(Z+\xi) r_0^2 \frac{d(h\nu)}{h\nu} \left\{\left[1 + \left(\frac{E'}{E}\right)^2 - \frac{2}{3}\frac{E'}{E}\right]\left[\ln 183\, Z^{-1/3} + \frac{1}{9}\frac{E'}{E}\right]\right\} \quad (3.29)$$

where*

$$\xi = Z\left(\Phi_{rad}^e/\Phi_{rad}^n\right) .$$

---

*The term ξ for most materials is a small correction. The latest estimates indicate $0.88 < \xi < 1.04$ for materials between Pb and Mg. Therefore $\xi = 1$ is good to a first approximation.[4]

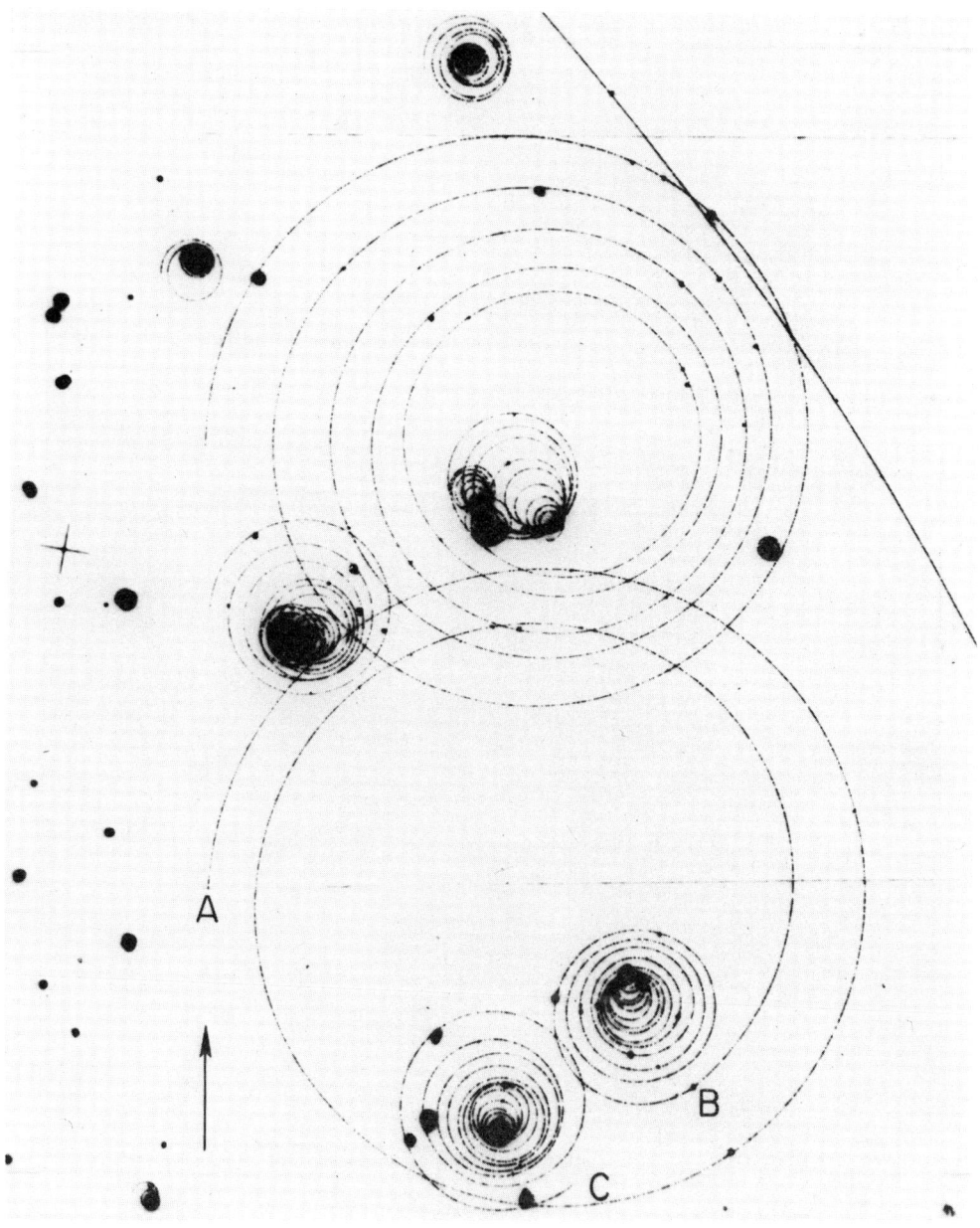

FIG. 3.2

Bremsstrahlung: The incident photon beam direction is indicated by the arrow. The Compton interaction at A produces an electron which loses a large fraction of its energy by radiation at B. The bremsstrahlung photon probably undergoes a Compton interaction at C.

Radiative energy loss by an electron is clearly shown in Fig. 3.2. The sudden increase in curvature of the incident electron path (under the influence of a magnetic field) indicates a large energy loss. The bremsstrahlung photon emitted does not leave a track but apparently makes a Compton interaction.

3.10 Radiative Energy Loss and the Radiation Length

The radiative energy loss of an electron passing through matter can be calculated from the probabilities stated in the previous section. Thus the energy lost by radiation is:

$$\left.\frac{dT}{dx}\right)_{rad} = \int_0^T h\nu \, \Phi_{rad}(T, h\nu) \, d(h\nu) \quad (\text{MeV-cm}^2\text{-g}^{-1}) \, .$$

If we neglect radiation in the field of electrons $\left(\text{i.e., } \Phi_{rad} = \Phi_{rad}^n\right)$, we get for the case of <u>no screening</u> ($m \ll E \ll 137 \, mZ^{-1/3}$)

$$\left.\frac{dT}{dx}\right)_{rad} = 4\alpha \, \frac{N_0}{A} \, Z^2 \, r_0^2 \, E \, \ln\left(\frac{2E}{m} - \frac{1}{3}\right) \quad (\text{MeV-cm}^2\text{-g}^{-1}) \tag{3.30}$$

and for <u>complete screening</u> ($E \gg 137 \, mZ^{-1/3}$)

$$\left.\frac{dT}{dx}\right)_{rad} = 4\alpha \, \frac{N_0}{A} \, Z^2 \, r_0^2 \, T \left[\ln(183 Z^{-1/3}) + \frac{1}{18}\right] \quad (\text{MeV-cm}^2\text{-g}^{-1}) \tag{3.31}$$

Note: $T \simeq E$.

It is convenient at this point to introduce the concept of <u>radiation length</u>. From Eq. (3.31) above it can be seen that at high energies

$$\frac{dT}{T} = -K \, dx$$

(we have now included the minus sign to indicate loss). Thus:

$$\frac{T(x)}{T(0)} = e^{-Kx}$$

where K is a constant for any given absorber. Consequently, the radiative energy loss will decrease exponentially with distance in the absorber. The distance over which the incident electron kinetic energy is reduced by a factor $1/e$ (due to radiative losses only) is defined as a <u>radiation length</u> and is denoted by X_0.* Hence when:

$$T(x)/T(0) = e^{-1}$$

$$Kx = 1$$

and

$$x = \frac{1}{K} \equiv X_0.$$

In the Bethe-Heitler formulation then (from Eq. (3.31)),

$$\frac{1}{X_0} = 4\alpha \frac{N_0}{A} Z^2 r_0^2 \left[\ln(183\, Z^{-1/3}) + \frac{1}{18} \right] \quad (cm^2\text{-}g^{-1}). \tag{3.32}$$

It can be seen that in the energy region where the concept of radiation length is valid (energy losses due primarily to radiative processes), $1/X_0$ is proportional to Z^2 and is independent of energy.

If we include the effect of atomic electrons and a correction for the Born approximation we get:[5]

$$\frac{1}{X_0} = \frac{4\alpha \frac{N_0}{A} Z(Z+1) r_0^2 \ln(183\, Z^{-1/3})}{1 + 0.12 \left(\frac{Z}{82}\right)^2} \quad (cm^2\text{-}g^{-1}) \tag{3.33}$$

3.11 Comparison of Collision and Radiative Energy Losses for Electrons

Comparison of the energy loss equations for collision processes with those for the radiative processes shows first that while collision energy loss increases with

---

*Dovzhenko and Pomanskii[9] derive, in accordance with current theoretical and experimental ideas, values for the radiation lengths and the critical energies of common materials.

Z, radiative energy loss increases with Z^2. Secondly, collision losses increase with lnE (for T > m) while radiative losses increase with E. Therefore at high energies, the radiation energy loss predominates. As the electron energy decreases, collision energy losses become significant until at a certain energy the two are equal. Below this energy collision losses predominate. This energy is called the <u>critical energy</u>, ϵ_0.

This critical energy can be approximated by[4]

$$\epsilon_0 = \left(\frac{800}{Z+1.2}\right) \text{MeV} \tag{3.34}$$

The ratio of radiative to collision energy loss is given approximately by (FBM):

$$\frac{(dT/dx)_{rad}}{(dT/dx)_{col}} \approx \frac{TZ}{800} \tag{3.35}$$

It is instructive also to consider the behavior of the fractional energy loss per radiation length for both processes (see Fig. 3.3).

For collision energy losses:

$$-\frac{1}{E}\left(\frac{dT}{dt}\right)_{col} \approx \frac{\ln E}{E Z}$$

where

$$t = x/X_0 \ .$$

For radiative energy losses:

at low energies ($\gamma \gg 1$)

$$-\frac{1}{E}\left(\frac{dT}{dt}\right)_{rad} = \frac{\ln\left(\frac{2E}{m} - \frac{1}{3}\right)}{\ln(183\, Z^{-1/3}) + \frac{1}{18}}$$

at high energies ($\gamma \approx 0$)

$$-\frac{1}{E}\left(\frac{dT}{dt}\right)_{rad} = \frac{T}{E}$$

This shows that at very high energies (> 1 GeV) where virtually all the energy losses are due to radiative processes the fractional energy loss per radiation length is independent of absorbing material and particle energy, and in fact is almost identical to 1 as shown in Fig. 3.3. Thus:

$$\frac{dT}{T} = - dt$$

which leads to

$$\frac{T(x)}{T(0)} = e^{-t} = e^{-x/X_0}$$

as we would expect.

It is apparent from Fig. 3.3 that the description of radiation phenomena is only slightly dependent on atomic number when thicknesses are measured in radiation lengths, and this dependence becomes less pronounced with increasing energy. Now, we have demonstrated in Chapter 2, by means of the Feynman diagram, that pair production is the photon interaction that is complementary to bremsstrahlung. Therefore, if in analytic shower theory the approximation is made that only pair production and bremsstrahlung interactions are important, one can expect that the longitudinal development of an electromagnetic cascade shower will be essentially Z-independent whenever the distance is expressed in radiation length units. This high energy approximation is commonly referred to as Approximation A in shower theory.[5]

3.12 Radiation Energy Losses by Heavy Particles

Without going into the details of heavy particle radiation loss probabilities, a classical treatment of the radiation loss process will show why these losses are generally negligible for heavy charged particles. Consider a particle of charge e, mass M and velocity β moving past a nucleus of charge Ze, and let $(1 - \beta) \ll 1$ (i.e., $\beta \approx 1$). If we consider the nucleus a point charge and assume its mass is

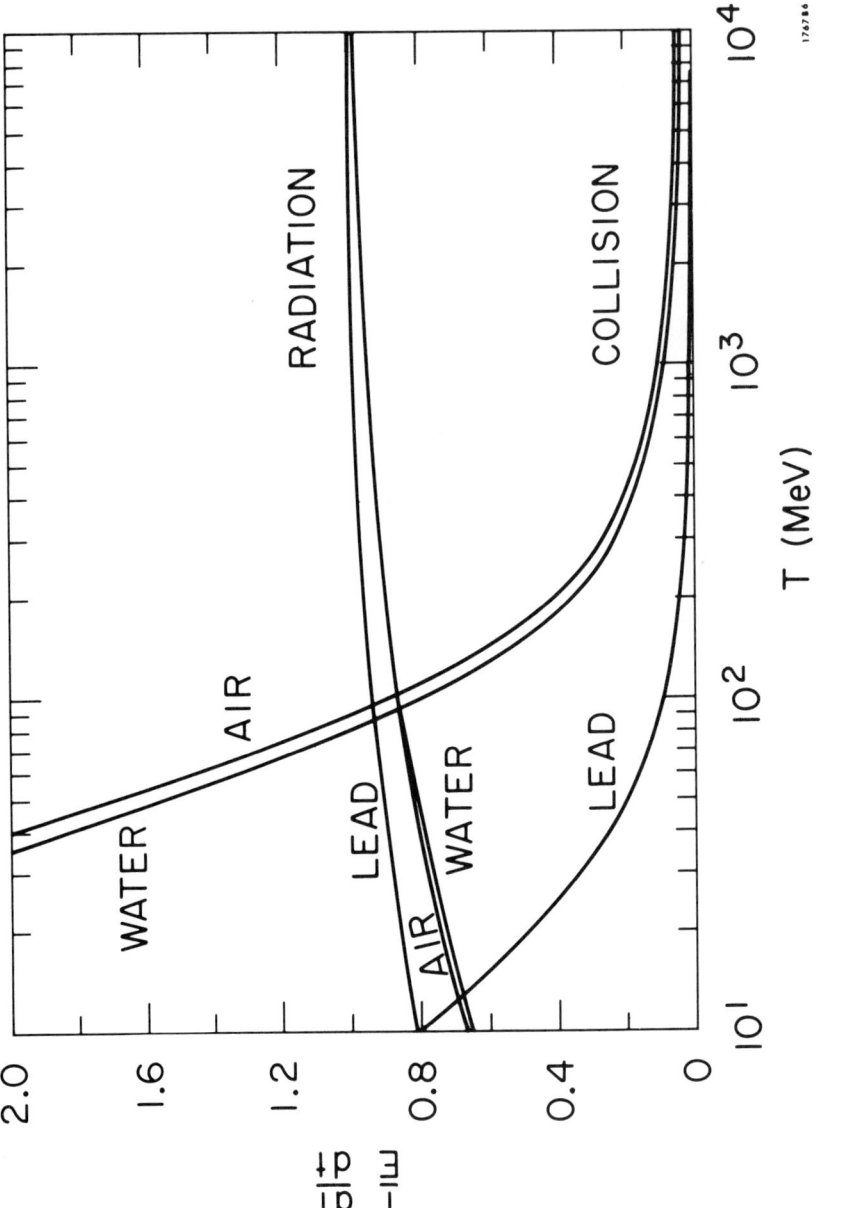

FIG. 3.3 Electron fractional energy loss per radiation length.

large compared with M, we can neglect any motion of the nucleus during the interactions. In the proximity of the nucleus the moving particle will be acted upon by a force[5]

$$F = \frac{Ze^2}{b^2} \frac{1}{(1-\beta^2)} = F_\perp$$

where b is the impact parameter. Hence the particle will undergo a maximum acceleration

$$a_{max} = \frac{F}{M}\sqrt{1-\beta^2} = \frac{Ze^2}{Mb^2}\frac{1}{\sqrt{1-\beta^2}}$$

According to classical electrodynamics this acceleration will cause the particle to radiate energy where the energy radiated per unit time is given by

$$\frac{2e^2}{3}a^2 \approx e^2 a^2$$

From this one can see that the energy radiated will be proportional to a^2 and hence the differential radiation probability

$$\Phi_{rad}(T, h\nu)\, d(h\nu) \propto \frac{Z^2 e^4}{M^2}$$

Now substituting $r_0^2 = e^2/m$ (classical radius of electron) we see that

$$\Phi_{rad}(T, h\nu)\, d(h\nu) \propto Z^2 r_0^2 \left(\frac{m}{M}\right)^2 .$$

This shows clearly that radiation energy losses are inversely proportional to the square of the particle mass.

Consequently the radiative energy loss by any particle of mass M will be less than that of an electron by a factor of $(m/M)^2$.

This is what we would have expected, however, since the same relationship appears in the complementary process for photons — namely, pair production. We see that for muons, the next closest mass to the electron, that

$$\Phi_{rad})_\mu = \Phi_{rad})_e \times \left(\frac{0.511 \text{ MeV}}{106 \text{ MeV}}\right)^2 = \frac{1}{40,000} \Phi_{rad})_e$$

so that for dosimetry purposes, we can neglect radiation losses by heavy charged particles.

3.13 Fluctuations in the Energy Loss by Radiation

Up to this point we have assumed that the radiative energy loss is continuous as an electron passes through an absorber. Consequently the formulas given (Eqs. (3.30) and (3.31)) are for average energy loss by radiation. However, the probability is significant that an electron loses a large fraction of its energy in a single radiative process. Therefore, we expect to find a distribution about the average for radiative energy loss just as we did for ionization loss. The corresponding probability function is:[5]

$$\omega(T_0, T, t)dT = \frac{dT}{T_0} \frac{[\ln(T_0/T)]^{(t/\ln 2)-1}}{\Gamma(t/\ln 2)}$$

where

$$\Gamma(t/\ln 2) = \int_0^\infty e^{-x} x^{(t/\ln 2)-1} dx$$

This distribution is significant when the radiative energy loss process predominates (i.e., $T > \epsilon_0$). In this energy region other processes become significant, namely cascade shower production. Consequently, an average radiative stopping power is no longer valid. A detailed treatment of radiative energy loss fluctuations will not be undertaken at this point. Analytic shower theory is discussed in detail in the text by Rossi.[5]

3.14 Range and Range Straggling

Since heavy charged particles or low energy electrons lose energy more or less continuously as they move through an absorber, they have a definite range.

This range can be calculated knowing the rate of energy loss. Consequently, the mean range R_0 of a particle of kinetic energy T is defined by:

$$R_0(T) = \int_0^T dT/(-dT/dx)$$

where $-\frac{dT}{dx}$ is given by the appropriate stopping power formula. This formula ignores mutliple scattering.

Now, the rate of energy loss is not strictly continuous but includes some statistical fluctuations as discussed previously. Therefore, there will be a distribution of ranges about the mean corresponding to the statistical distribution of energy loss. Since the energy loss process is Gaussian for thick absorbers, the range distribution is also Gaussian. The probability P(R)dR of a particle with an initial energy T having a range between R and R + dR is given by[3]

$$P(R)dR = \frac{1}{\sigma\sqrt{2\pi}} \exp\left[-\frac{(R-R_0)^2}{2\sigma^2}\right] dR \qquad (3.36)$$

where

$$\sigma^2 \equiv \langle (R-R_0)^2 \rangle_{av} = \int_{-\infty}^{\infty} P(R)(R-R_0)^2 dR$$

The quantity $\langle (R-R_0)^2 \rangle_{av}$ is generally obtained from measurements of the number of particles penetrating to a given distance. Because of the Gaussian nature of the distribution the relative number-distance curve is as shown in Fig. 3.4:

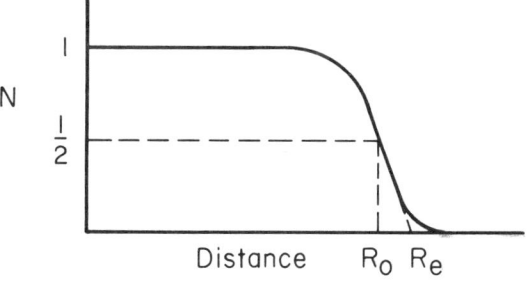

FIG. 3.4

The point R_0 where the curve of N has one-half its maximum value is also the point at which the curve has its maximum slope, $-1/\sigma\sqrt{2\pi}$. By constructing a tangent to the curve at this point and extrapolating to the R-axis intersection one obtains the point R_e known as the extrapolated range. The relationship between R_e and the mean range R_0 is given by the equation for the tangent line:

$$(y_2 - y_1) = m(x_2 - x_1)$$
$$1/2 - 0 = -1/\sigma\sqrt{2\pi} \ (R_0 - R_e)$$

or

$$R_e - R_0 = \frac{\sigma\sqrt{2\pi}}{2} = S$$

where S is defined as the <u>straggling</u> parameter and

$$S^2 = \frac{\sigma^2 \pi}{2} = \frac{\pi}{2} \langle (R - R_0)^2 \rangle_{av}$$

The percentage straggling is defined as

$$\epsilon \equiv \frac{100\sqrt{\langle (R - R_0)^2 \rangle_{av}}}{R_0}$$

$$= \frac{100 S}{R_0} \sqrt{\frac{2}{\pi}}$$

The percentage straggling decreases slowly as the initial particle energy increases until a minimum is reached at $T/M = 2.5$. It may be recalled that this is the same region at which the minimum is reached in the stopping power curve. Beyond this minimum, ϵ again increases reflecting the influence of the $(1-\beta^2)^{-1}$ term. It turns out that ϵ also increases slowly with Z, varying about 25% from beryllium to lead.

This treatment of particle range is not applicable to high energy electrons where the predominant energy losses result from the production of bremsstrahlung. When the electron energy is above the critical energy for the absorbing material, one should use a mean rate of energy loss due to collisions and bremsstrahlung in the above definition of range.[10] For energies much larger than the critical energy, the concept of electron range is meaningless because of cascade shower production.

3.15 Elastic Scattering of Charged Particles

When a charged particle passes in the neighborhood of a nucleus, it undergoes a change in direction, referred to as scattering. Because of the relatively small probability that a photon is emitted with energy comparable to the kinetic energy of the charged particle, the scattering process is generally considered to be an elastic one. In addition we assume that the nucleus is very much heavier than the incident particle and thus does not acquire significant kinetic energy.

We define the differential scattering probability as follows:

$\Xi(\theta)d\omega\, dx$ = probability that a charged particle of momentum p and velocity β, traversing a thickness $dx(g\text{-}cm^{-2})$, undergoes a collision which deflects the trajectory of the particle into the solid angle $d\omega$ about θ (from its original direction).

Various formulas have been derived for $\Xi(\theta)d\omega\, dx$, which depends on the nature of the medium as well as the charge and spin of the particle. If we neglect the shielding of a point charge, Ze, by the atomic electrons, and if we use the Born approximation, we can obtain the following expressions for heavy singly charged particles (c = 1 units).[5]

A. Spin Zero Particles (e.g., alpha particles and pions)

$$\Xi(\theta)d\omega = \frac{1}{4} N_0 \frac{Z^2}{A} r_0^2 \left(\frac{m}{p\beta}\right)^2 \frac{d\omega}{\sin^4(\theta/2)} \quad (cm^2\text{-}g^{-1}) \qquad (3.37)$$

where

N_0 = Avogadro's number

m = mass of electron

$r_0 = e^2/m$ = classical electron radius

Z = atomic number

A = atomic weight

Note: for alpha-particles, multiply by $z^2 = (2)^2 = 4$

B. Spin One-Half Particles (e.g., protons and muons)

$$\Xi(\theta)d\omega = \frac{1}{4}N_0 \frac{Z^2}{A} r_0^2 \left(\frac{m}{p\beta}\right)^2 \frac{d\omega}{\sin^4(\theta/2)} (1 - \beta^2 \sin^2(\theta/2))(cm^2\text{-}g^{-1}) \quad (3.38)$$

This formula is called the Mott scattering formula for heavy particles.

C. Electron Scattering

Mott[11] derived the elastic scattering cross section for electron scattering from nuclei of charge Ze by employing the relativistic Dirac theory with the Born approximation. By expanding Mott's exact formula in powers of αZ, McKinley and Feshbach[12] obtained

$$\Xi(\theta)d\omega = \frac{1}{4}N_0 \frac{Z^2}{A} r_0^2 \left(\frac{m}{p\beta}\right)^2 \frac{d\omega}{\sin^4(\theta/2)}$$

$$\times \left[1 - \beta^2 \sin^2(\theta/2) + \pi\beta\alpha Z(1 - \sin(\theta/2))\sin(\theta/2)\right] (cm^2\text{-}g^{-1}) \quad (3.39)$$

where $\alpha = 1/137 =$ fine structure constant.

Note: The above formula is valid only for high velocities ($\beta \sim 1$) and for rather low Z materials ($\alpha Z \leq 0.2 \rightarrow Z \lesssim 27$)

D. Rutherford Scattering Formula

For small deflections, $\sin(\theta/2) \sim \theta/2$, and we can neglect the spin terms, so that all formulas above become

$$\Xi(\theta)d\omega = 4N_0 \frac{Z^2}{A} r_0^2 \left(\frac{m}{p\beta}\right)^2 \frac{d\omega}{\theta^4} (cm^2\text{-}g^{-1}) \quad (3.40)$$

This equation, as well as the previous ones, are not defined at $\theta = 0$.*

E. Derivation of the Rutherford Scattering Formula Using the Born Approximation

The basic formula for the Born approximation is given by[13]

$$\frac{d\sigma}{d\omega} d\omega = \frac{p^2}{4\pi^2 \hbar^4 v^2} |\langle k'|U|k\rangle|^2 d\omega \quad (cm^2/atom) \quad (3.41)$$

---

*Note: We will see shortly that $\theta > \theta_1$ (Eq. (3.43)) due to electron screening.

where $\bar{p}$ = particle momentum = $\hbar\bar{k}$

  v = particle velocity

and

$$\langle k' | U | k \rangle = \int_{\text{allspace}} \exp[-i\bar{k}' \cdot \bar{r}] U(r) \exp[i\bar{k} \cdot \bar{r}] d^3r$$

= matrix element between the initial state ($\bar{k}$) and final state ($\bar{k}'$).

(Essentially, the Born approximation comes from Fermi's Golden Rule No. 2 with the approximations:

1) $\psi_i \sim e^{i\bar{k} \cdot \bar{r}}$ plane wave incident
2) $\psi_f \sim e^{i\bar{k}' \cdot \bar{r}}$ plane wave out
3) H = Hamiltonian of the interaction = $U(r)$ only
4) Fixed point scattering center.)

Now, the scattering process is described by the diagrams,

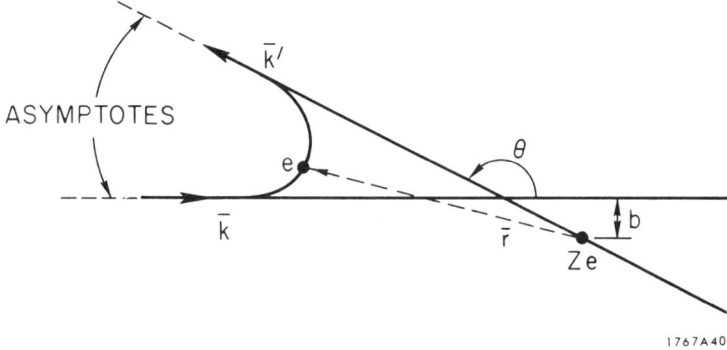

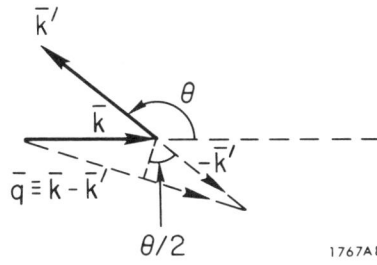

so that $k = k'$ (i.e., **elastic** scattering) and $q = |\bar{k} - \bar{k'}| = 2k \sin(\vartheta/2)$. Let

$$U(r) = \frac{Ze^2}{r} e^{-r/r_a}$$

and

$$(\bar{k} - \bar{k'}) \cdot \bar{r} = \bar{q} \cdot \bar{r} = qr\mu$$

where

$$\mu \equiv \cos \sphericalangle (\bar{q}, \bar{r}) \equiv \cos \beta$$

and

$$r_a = r_0 \, Z^{-1/3}/\alpha^2 = \text{radius of atom (Fermi-Thomas model}^{5,22})$$

Furthermore

$$d^3r = r^2 \sin \beta \, dr \, d\beta \, d\phi = 2\pi r^2 \, dr \, d\mu$$

(neglecting the sign and assuming azimuthal symmetry). So that

$$\langle k'|U|k \rangle = \int e^{i(\bar{k}-\bar{k'}) \cdot \bar{r}} \, U(r) \, d^3r$$

$$= 2\pi \, Ze^2 \int_0^\infty \int_{-1}^1 \frac{e^{iqr\mu}}{r} e^{-r/r_a} r^2 \, dr \, d\mu$$

$$= \frac{4\pi \, Ze^2}{q^2} \int_0^\infty e^{-\xi x} \sin x \, dx$$

where $x \equiv qr$ and $\xi = (qr_a)^{-1}$. Now, one can integrate by parts twice to obtain

$$\int_0^\infty e^{-\xi x} \sin x \, dx = -\left(\frac{1}{1+\xi^2}\right)\left[(\cos x + \xi \sin x) e^{-\xi x}\right]\Big|_0^\infty$$

$$= \frac{1}{1+\xi^2} = \frac{1}{1+(qr_a)^{-2}}$$

Therefore,

$$\langle k'|U|k\rangle = \frac{4\pi Z e^2}{q^2} \frac{1}{1+(qr_a)^{-2}}$$

And from Eq. (3.41)

$$\frac{d\sigma}{d\omega} d\omega = \frac{p^2}{4\pi^2 \hbar^4 v^2} |\langle k'|U|k\rangle|^2 d\omega$$

$$= \frac{p^2}{4\pi^2 \hbar^4 v^2} \frac{16\pi^2 Z^2 e^4}{q^4 \left[1+(qr_a)^{-2}\right]^2}$$

Now,

$$p = \hbar k = \hbar/\lambdabar \qquad \text{(de Broglie wavelength)}$$

$$\beta = v \qquad \text{(c = 1 units)}$$

$$r_0 = e^2/m \qquad \text{(c = 1 units)}$$

and we can define θ_1 by

$$\theta_1 = \alpha^2 Z^{1/3} \lambdabar/r_0$$

so that

$$qr_a = [2k\sin(\theta/2)] \frac{r_0 Z^{-1/3}}{\alpha^2} = \left[\frac{2}{\lambdabar}\sin(\theta/2)\right]\frac{\lambdabar}{\theta_1}$$

$$= \frac{2}{\theta_1}\sin(\theta/2)$$

Therefore,

$$\frac{d\sigma}{d\omega} d\omega = \frac{1}{4} Z^2 r_0^2 \left(\frac{m}{p\beta}\right)^2 \frac{d\omega}{\left[\sin^2(\theta/2) + \frac{1}{4}\theta_1^2\right]^2}$$

$$\simeq 4 Z^2 r_0^2 \left(\frac{m}{p\beta}\right)^2 \frac{d\omega}{\left[\theta^2 + \theta_1^2\right]^2} \quad (\text{cm}^2/\text{atom})$$

for small angles.

And,

$$\Xi(\theta) \, d\omega = \frac{N_0}{A} \frac{d\sigma}{d\omega} d\omega = \text{probability per g - cm}^2$$

$$= 4 N_0 \frac{Z^2}{A} r_0^2 \left(\frac{m}{p\beta}\right)^2 \frac{d\omega}{\left[\theta^2 + \theta_1^2\right]^2} \quad (3.42)$$

which is a form of the Rutherford scattering formula.

The θ_1 term accounts for the screening of the electric field of the nucleus by the outer electrons.[14,15]

At this point, a few general remarks are in order. First of all, when a charged particle penetrates an absorbing medium, most of the scattering interactions lead to very small deflections. Small net deflections are generally the result of a large number of very small deflections; whereas, large net deflections are the result of a single large-angle scatter plus a number of very small deflections. Because of this fact, one refers to the small-angle scattering as multiple scattering and the large-angle scattering is called single scattering. The intermediate case is known as plural scattering.

Secondly, one can compute the scattering probability in the field of the atomic electrons to obtain

$$\Xi(\theta) \, d\omega \simeq 4 N_0 \frac{Z r_0^2}{A} \left(\frac{m}{p\beta}\right)^2 \frac{d\omega}{\theta^4}$$

Hence, even though collisions with atomic electrons are responsible for almost all of the energy loss, their contribution to scattering is fairly small (10% for $Z = 10$, 1% for $Z = 82$).

It should be noted that the actual process is complicated by the fact that the scattering from atomic electrons is inelastic whereas the above formula is for an elastic process. Furthermore, it can only apply to heavy charged particles since electron-electron interactions must account for exchange effects (i.e., identical particles require one to invoke the Pauli exclusion principle). The net result is the same since Z^2 is usually replaced by $Z(Z + 1)$ in the cross sections given so far (similar to the corrections made in the radiative probabilities).

Finally, the expressions that have been presented have been derived under the assumption of a point charge, Ze. The finite size of the nucleus, as well as the screening of its field by the atomic electrons, limit the validity of the results to a certain range of angular deflections. The effect of screening has been studied both by Goudsmit and Saunderson[14] and by Moliere.[15] According to Rossi,[5] the screening of the electric field of the nucleus by the outer electrons does not appreciable affect the scattering probability until

$$\theta_1 \simeq \alpha Z^{1/3} (m/p) = \alpha^2 Z^{1/3} \lambda/r_0 \qquad (3.43)$$

This quantity shows up in the Goudsmit and Saunderson calculation and in the derivation above as follows (for small angles):

$$\Xi(\theta) \, d\omega = 4 N_0 \frac{Z^2}{A} r_0^2 \left(\frac{m}{p\beta}\right)^2 \frac{d\omega}{\left(\theta^2 + \theta_1^2\right)^2}$$

Note that $\Xi(\theta) \, d\omega$ no longer diverges as $\theta \to 0$.

In order to account for the finite size of the nucleus, Williams[16] finds that the range of validity of the formulas for $\Xi(\theta)d\omega$ is limited (for large angles) by

$$\theta_2 \simeq 280 \, A^{-1/3} \, (m/p) \tag{3.44}$$

We will make use of these limits in the next section.

F. The Mean Square Angle of Scattering

Assume that a charged particle traverses a medium of finite thickness $x(g\text{-}cm^{-2})$. The value of $\langle\theta^2\rangle$ at $x+dx$ equals the value of $\langle\theta^2\rangle$ at x plus the mean square angle of scattering in dx which is[5]

$$d\langle\theta^2\rangle = \int_\theta \theta^2 \, \Xi(\theta)d\omega \, dx$$

This may be rewritten as:

$$\frac{d\langle\theta^2\rangle}{dx} = \theta_s^2$$

where

$$\theta_s^2 = \int_\theta \theta^2 \, \Xi(\theta)d\omega$$

$$= 8\pi N_0 \frac{Z^2}{A} r_0^2 \left(\frac{m}{p\beta}\right)^2 \int_{\theta_1}^{\theta_2} d\theta/\theta$$

$$= 8\pi N_0 \frac{Z^2}{A} r_0^2 \left(\frac{m}{p\beta}\right)^2 \ln(\theta_2/\theta_1)$$

$$= 16\pi N_0 \frac{Z^2}{A} r_0^2 \left(\frac{m}{p\beta}\right)^2 \ln\left[196 \, (Z/A)^{1/6} \, Z^{-1/3}\right]$$

where we have assumed that

a. $\Xi(\theta)d\omega$ given by the Rutherford scattering formula (3.40),

b. the charged particle undergoes a large number of very small angle collisions, so that $\sin\theta \sim \theta$,

c. $\Xi(\theta) = 0$ for $\theta < \theta_1$ or $\theta > \theta_2$,

d. θ_1 and θ_2 given by Eqs. (3.43) and (3.44), respectively.

Now, the coefficient $196 \, (Z/A)^{1/6}$ in the logarithm varies from 175 to 169 for $A = 2Z$ (low Z) and $A = 2.5 \, Z$ (high Z) respectively. Furthermore, from the definition of the radiation length (see Sec. 3.10),

$$\frac{1}{X_0} = 4\alpha \frac{N_0}{A} Z^2 r_0^2 \ln[183 Z^{-1/3}] \quad (cm^2-g^{-1})$$

so that we can, to a good approximation, reexpress θ_s in units of reciprocal radiation lengths, to obtain

$$\theta_s^2 = \left(\frac{E_s}{p\beta}\right)^2 \frac{1}{X_0} \tag{3.45}$$

where E_s is defined by

$$E_s = \left(\frac{4\pi}{\alpha}\right)^{1/2} m = 21.2 \text{ MeV} \tag{3.46}$$

Thus,

$$d\langle\theta^2\rangle = \left(\frac{E_s}{p\beta}\right)^2 dx \quad \text{(square radians)}$$

where dx is now expressed in radiation lengths.

If the scattering layer is sufficiently thin so that energy loss can be neglected, then θ_s^2 is constant, and we find that

$$\theta_{rms} = \frac{21\sqrt{x}}{p\beta} \tag{3.47}$$

where $p\beta$ is in MeV/c, x is radiation lengths, and θ_{rms} is in radians. For high energy electrons,

$$p\beta = \beta^2 E = \beta^2(T+m) \simeq T$$

to a very good approximation. We then find that

$$\theta_{rms} = \frac{21\sqrt{x}}{T} \tag{3.48}$$

Often it is more convenient to consider the projected angle, θ_y rather than the total (space) angle θ. It can be shown that[5]

$$\langle\theta_y^2\rangle = \frac{1}{2}\langle\theta^2\rangle \tag{3.49}$$

so that

$$(\theta_y)_{rms} = \frac{15\sqrt{x}}{p\beta} \qquad (3.50)$$

G. Fermi-Eyges Theory of Multiple Scattering with Energy Loss

The analytical treatment of this process is very difficult unless one makes extensive approximations. A review of the various approaches has been given by Zerby and Keller.[17] One of the most widely used computations is attributed to Fermi and Eyges, and is also called Gaussian scattering. We will briefly discuss this treatment since the results are quite often used in shielding calculations around high energy accelerators.[18]

The basic equation is the Fermi diffusion equation (FDE):

$$\frac{\partial F(x, y, \theta_y)}{\partial x} = -\theta_y \frac{\partial F}{\partial y} + \frac{1}{W^2} \frac{\partial^2 F}{\partial \theta_y^2} \qquad (3.51)$$

where

$$W = 2p\beta/E_s$$

and where $F(x, y, \theta_y) dy\, d\theta_y$ = number of particles at x having lateral displacement (y, dy) and traveling at an angle $(\theta_y, d\theta_y)$

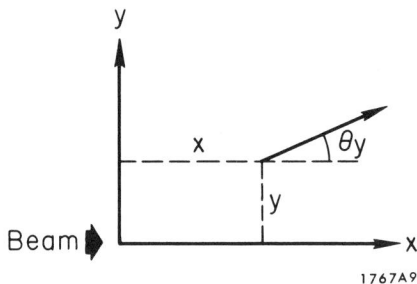

The derivation of this equation is given, for example, by Barkas.[19]

Eyges[20] solves this equation (with energy loss included) by the method of Fourier transforms, with the assumption that p, β, and hence W, are functions of x — that is, the fact that a particle at x has traveled a somewhat greater distance than x, due to the deviations caused by scattering, is neglected (a good approximation for high energy particles).

The final result of Eyges' mathematics is

$$F(x, y, \theta_y) = \frac{1}{4\pi \sqrt{B}} \exp\left[-\frac{\theta_y^2 A_2 - 2y\theta_y A_1 + y^2 A_0}{4B}\right] \quad (3.52)$$

where

$$\left.\begin{array}{l} B(x) = A_0 A_2 - A_1^2 \\[6pt] A_0(x) = \displaystyle\int_0^x \frac{d\eta}{W^2(\eta)} \\[10pt] A_1(x) = \displaystyle\int_0^x \frac{(x-\eta)\,d\eta}{W^2(\eta)} \\[10pt] A_2(x) = \displaystyle\int_0^x \frac{(x-\eta)^2\,d\eta}{W^2(\eta)} \end{array}\right\} \quad (3.53)$$

Now, if we integrate over θ_y, we obtain the lateral distribution — independent of angle:

$$H(x, y)dy = \int_{\theta_y} F(x, y, \theta_y) d\theta_y \, dy$$

$$= \frac{1}{2\sqrt{\pi A_2}} e^{-y^2/4A_2} \, dy \quad (3.54)$$

Similarly, the angular distribution — irrespective of displacement — is:

$$G(x, \theta_y)d\theta_y = \int_y F(x, y, \theta_y) dy \, d\theta_y$$

$$= \frac{1}{2\sqrt{\pi A_0}} e^{-\theta_y^2/4A_0} \, d\theta_y \quad (3.55)$$

Thus, we get Gaussian distributions for G and H as anticipated.

The mean square projected angle of scattering is defined by

$$\langle \theta_y^2 \rangle = \int \theta_y^2 G(x, \theta_y) d\theta_y = 2A_0(x) = 2 \int_0^x \frac{d\eta}{W^2(\eta)}$$

If we assume that the scattering medium is thick enough to consider the scattering to be essentially multiple, but thin enough to neglect energy loss, we have

$$W = 2p\beta/E_s = \text{constant}$$

and therefore

$$\langle \theta_y^2 \rangle = 2x/W^2$$

or

$$(\theta_y)_{rms} = \frac{15\sqrt{x}}{p\beta}$$

as before (see Eq. (3.50)).

A more complete treatment of multiple scattering, which allows for plural and single scattering as well, has been done by Moliere,[15] and Scott,[21] and is beyond the present discussion.

3.16 Scaling Laws for Stopping Power and Range

As we have shown in Section 3.4, the unrestricted mass stopping power for a heavy charged particle of mass M, charge z, and velocity β traveling in a medium of atomic number Z, atomic weight A, and density ρ is of the form

$$\frac{1}{\rho}\frac{dT}{dx}\bigg)_{col} \propto \left(\frac{Z}{A}\right)(z^2 f(\beta, I)) \, (\text{MeV-cm}^2\text{-g}^{-1}) \qquad (3.56)$$

where

$$f(\beta, I) = \beta^{-2} \left[\ln\left(\frac{2m\beta^2}{(1-\beta^2)I}\right) - 2\beta^2 \right]$$

and where the specific dependence on density is now indicated. For two particles, 1 and 2, of different charge but moving with the same velocity in a given medium, we have

$$\frac{\frac{1}{\rho}\left(\frac{dT}{dx}\right)^1_{col}}{\frac{1}{\rho}\left(\frac{dT}{dx}\right)^2_{col}} = \left(\frac{z_1}{z_2}\right)^2 \qquad (3.57)$$

independent of their individual masses.

Now, to a first approximation the logarithmic term is a weak function of I. Hence for a charged particle of velocity β traveling in two different media, a and b we have

$$\frac{\frac{1}{\rho}\left(\frac{dT}{dx}\right)^a_{col}}{\frac{1}{\rho}\left(\frac{dT}{dx}\right)^b_{col}} = \frac{\left(\frac{Z}{A}\right)_a}{\left(\frac{Z}{A}\right)_b} \qquad (3.58)$$

If we further assume that

$$\frac{Z}{A} \simeq \frac{1}{2}$$

Then

$$\frac{1}{\rho}\left(\frac{dT}{dx}\right)^a_{col} \simeq \frac{1}{\rho}\left(\frac{dT}{dx}\right)^b_{col} \qquad (3.59)$$

Thus, to a good approximation the only difference in the shielding power of various materials is due to their densities. This suggests that absorber thicknesses be measured in g/cm^2.

The range of a heavy charged particle is given by

$$R(cm) = \int_{T_0}^{0} \frac{dT}{(-dT/dx)}$$

Now,

$$T = (\gamma - 1)M$$

where

$$\gamma = (1 - \beta^2)^{-1/2}$$

so that

$$R \propto \frac{A}{\rho Z z^2} \int_0^{\beta_0} \frac{\beta^2}{\left[\ln\left[\frac{2m\beta^2}{I(1-\beta^2)}\right] - 2\beta^2\right]} \frac{M\beta \, d\beta}{(1-\beta^2)^{3/2}}$$

or

$$R \propto \left(\frac{M}{z^2}\right)\left(\frac{A}{\rho Z}\right) F(\beta_0, I) \tag{3.60}$$

Therefore, for two different particles, 1 and 2, traveling with the same velocity in a given medium, we have that

$$\frac{R_1}{R_2} = \left(\frac{M_1}{M_2}\right)\left(\frac{z_2}{z_1}\right)^2 \tag{3.61}$$

And if we make the reasonable assumption that $F(\beta_0, I)$ depends only weakly on I, then for the same particle traveling in two different media, a and b, we have

$$\frac{R_a}{R_b} = \frac{(Z/A)_b}{(Z/A)_a} \left(\frac{\rho_b}{\rho_a}\right) \tag{3.62}$$

and with the further approximation that $Z/A \simeq 1/2$, we have

$$\frac{R_a}{R_b} \simeq \rho_b/\rho_a \tag{3.63}$$

as we have previously indicated.

Finally, there is another convenient way to scale the unrestricted stopping power for heavy particles having the same charge. Consider Eq. (3.13) in the approximate form

$$\frac{1}{\rho}\left.\frac{dT}{dx}\right)_{col} \simeq \frac{4Cm}{\beta^2} \ln\left[\frac{2m\beta^2}{I(1-\beta^2)}\right] \tag{3.64}$$

From the relativistic equations

$$E = \gamma M = T + M$$

and

$$\gamma^2 = 1/1-\beta^2$$

so that Eq. (3.64) becomes

$$\frac{1}{\rho}\frac{dT}{dx}\bigg)_{col} \cong \frac{4\,Cm\left(\frac{T}{M}+1\right)^2}{\frac{T}{M}\left(\frac{T}{M}+2\right)} \ln\left[\left(\frac{2m}{I}\right)\left(\frac{T}{M}\right)\left(\frac{T}{M}+2\right)\right]$$

which suggests that the stopping power curve for different particles of the same z will be essentially the same when plotted against the ratio T/M. This is illustrated in Fig. 3.5 using the stopping power data of Barkas and Berger.[1] Notice that the minimum occurs at about T/M = 3, as previously indicated in Section 3.4.

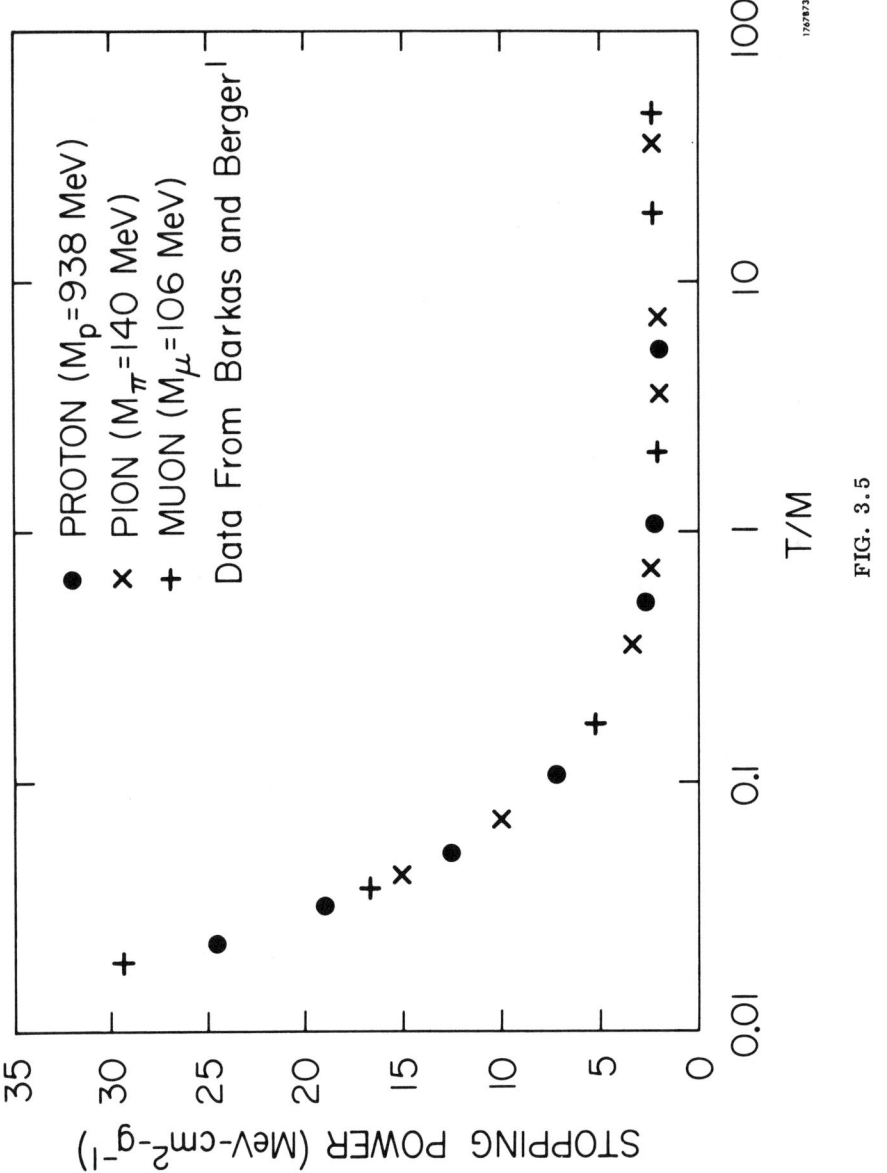

FIG. 3.5 Mass stopping power for heavy charged particles.

REFERENCES

1. W. H. Barkas and M. J. Berger, Tables of Energy Losses and Ranges of Heavy Charged Particles, (National Aeronautics and Space Administration, Washington, D.C., 1964), NASA-SP-3013.

2. H. A. Bethe, Ann. Physik $\underline{5}$, 325 (1930); Z. Physik $\underline{76}$, 293 (1932).

3. R. M. Sternheimer, Phys. Rev. $\underline{88}$, 851 (1952); $\underline{103}$, 511 (1956); $\underline{145}$, 247 (1966); Methods of Experimental Physics, Vol. 5 - Part A, Chapter 1, L.C.L. Yuan and C. S. Wu, Editors, (Academic Press, New York, 1961).

4. M. J. Berger and S. M. Seltzer, Tables of Energy Losses and Ranges of Electrons and Positrons, (National Aeronautics and Space Administration, Washington, D.C., 1964), NASA-SP-3012.

5. B. Rossi, High Energy Particles, (Prentice Hall, Inc., Englewood Cliffs, New Jersey, 1952).

6. L. D. Landau, J. Phys. USSR $\underline{8}$, 201 (1944).

7. K. R. Symon, Harvard University, Thesis (1948).

8. H. A. Bethe and W. Heitler, Proc. Roy. Soc. $\underline{A146}$, 83 (1934).

9. O. I. Dovzhenko and A. A. Pomanskii, Soviet Physics JETP $\underline{18}$, 187 (1964).

10. M. J. Berger and S. M. Seltzer, Phys. Rev. $\underline{C2}$, 621 (1970).

11. N. F. Mott, Proc. Roy. Soc. (London) $\underline{A124}$, 425 (1929); $\underline{A135}$, 429 (1932).

12. W. A. McKinley, Jr., and H. Feshback, Phys. Rev. $\underline{74}$, 1759 (1948).

13. E. Segre, Nuclei and Particles, (W. A. Benjamin, Inc., New York, 1965).

14. S. A. Goudsmit and J. L. Saunderson, Phys. Rev. $\underline{57}$, 24 (1940); $\underline{58}$, 36 (1940).

15. G. Moliere, Z. Naturforschung $\underline{2a}$, 133 (1947); Z. Naturforschung $\underline{3a}$, 78 (1948).

16. E. J. Williams, Proc. Roy. Soc. $\underline{A169}$, 531 (1939).

17. C. D. Zerby and F. L. Keller, Nucl. Sci. and Engin. $\underline{27}$, 190 (1967).

18. W. R. Nelson, Nucl. Instr. and Methods $\underline{66}$, 293 (1968).

19. Walter H. Barkas, Nuclear Research Emulsions (Academic Press, New York, 1963).

20. L. Eyges, Phys. Rev. $\underline{74}$, 1534 (1948).

21. W. T. Scott, Rev. Mod. Phys. $\underline{35}$, 231 (1963).

22. L. I. Schiff, Quantum Mechanics (McGraw-Hill Book Co., Inc., New York, 1949), pp. 271-273.

23. H. D. Maccabee and D. G. Papworth, Phys. Lett. $\underline{30A}$, 241 (1969).

MAIN REFERENCE

(FBM)　J. J. Fitzgerald, G. L. Brownell, and F. J. Mahoney, Mathematical Theory of Radiation Dosimetry (Gordon and Breach, New York, 1967).

CHAPTER 4

ENERGY DISTRIBUTION IN MATTER

4.1 Introduction

The quantity "absorbed dose" as defined by the ICRU (see Section 1.2, items 4 and 5) is a macroscopic concept like other physical quantities such as temperature and pressure. It is useful because it specifies in a single number the energy concentration near the point of interest. However, it is obvious because of the differences in biological responses to equal absorbed doses of different radiations that local energy densities and microscopic distributions are important in some instances. For some radiations the local energy densities can be significantly different from the absorbed dose. (See Fig. 1.1.)

The local energy density is the quotient E/m where E is the energy deposited in a mass element m. Its symbol is Z and it has units of ergs/g. The difference between Z and the absorbed dose $D (= dE_D/dm)$ comes about as one shrinks the mass element about the point of interest. When the mass element becomes very small the energy losses of the charged particles passing through m are no longer averaged out and in fact Z will be zero in the majority of instances. When Z is not zero, moreover, it can be very much larger than D. These great fluctuations in Z come about because energy is lost by charged particles in discrete steps. Thus the local energy density in a small mass element will depend on the number of charged particles traversing the mass and the amount of energy each happens to lose during the traversal.

In this chapter we will discuss linear energy transfer (LET), LET distributions and energy density distributions. Although these processes are microscopic and somewhat peripheral to the calculation and measurement of absorbed dose, they are important in understanding the energy loss process and the relationships between absorbed dose and radiation effects.

4.2 Linear Energy Transfer

We have previously discussed (Chapters 2 and 3) the interactions of charged and uncharged particles with an absorbing medium. The deposition of energy in a medium is through the interactions of charged particles with the atoms of the absorber and the average rate of energy loss is given by the stopping power formula appropriate for the charged particle of interest. At low energies stopping power is an inverse function of the square of the particle velocity. Thus it is obvious that as the particle slows down, the rate of energy loss increases. Consequently a large amount of energy can be deposited in a small mass element. Some of the knock-on electrons set in motion through the charged particle interactions can have significant kinetic energy, however, and deposit some of this energy outside the mass element about the point of interest. This is called "delta ray" production.

The collision probability is given by (see Chapter 3)

$$\Phi_{col} \propto \frac{z^2}{\beta^2 (T')^2}$$

where

z is the charge of the moving particle

β is the particle velocity ($c = 1$ units)

T' is the energy transferred in the collision.

Consequently the probability for a collision to occur is higher for a slow particle transferring small amounts of energy in each collision. From Table 4.1 it is obvious that, based on velocity, the interaction probability for a given energy is much greater for a proton or alpha particle than for an electron. Therefore, the α-particle or proton has a much greater collision density for a given energy than an electron.

Table 4.1

β^2

	Particle		
Energy (MeV)	e⁻	p	α
1	.9	.002	.0004
10	1.0	.02	.004
100	1.0	.17	.06
1,000	1.0	.75	.36
10,000	1.0	.99	.9

Now consider the energy transferred in a single collision (T'). For the electron $T'_{max} = T/2$ while for the heavy particles $T'_{max} \simeq 2m\,(\beta^2/(1-\beta^2))$. From Table 4.2 we see, based on the maximum energy transferred, that the interaction probability again is much greater for the heavy particles than for the electron.

Table 4.2

T'_{max} (MeV)

Energy (MeV)	e⁻	p	α
1	.5	.002	.0004
10	5.0	.02	.004
100	50.0	.2	.065
1,000	500.0	3.0	.55
10,000	5,000.0	99.0	9.0

As an example, for particles of energy 1 MeV the collision probabilities would be in the approximate ratio:

$$\Phi^e_{col} : \Phi^p_{col} : \Phi^\alpha_{col} = 1 : 10^7 : 10^{10} \ .$$

The points to be made are that for a given kinetic energy T,

1) Heavy particles are more likely to interact than are electrons.
2) The energy transferred per interaction is much less for heavy particles. Consequently, δ-ray production is much less and the local energy deposition is much greater.

The linear energy transfer (LET) concept is a description of the rate of energy loss from the standpoint of the absorber. As such it considers only the energy "locally imparted" to the absorber. It is different from stopping power in that LET refers to the average rate of energy deposited in a limited volume whereas stopping power refers to the average rate of energy lost no matter where in the absorber it is deposited.* Thus in the LET concept an upper limit is placed on the discrete energy losses beyond which the losses are no longer considered local. As was pointed out in Section 3.5, we can take this upper limit as T'_{max} and calculate LET_∞ which is equal to the stopping power. Conversely we can also calculate a restricted stopping power corresponding to a maximum energy transfer less than T'_{max}.

4.3 Delta Rays

As we have discussed in Chapter 3, the principal mode of charged particle energy loss is through collisions for all particles except high energy electrons. These collisions can be classified into two types depending on the impact parameter.

---

*The term stopping power will imply unrestricted stopping power and corresponds numerically to LET_∞.

Distant (or soft) collisions are most probable and result in small energy transfers. Near (or hard) collisions on the other hand can transfer a large amount of energy to the secondary electron. The amount of energy transferred is governed by the collision kinematics discussed in Chapter 3. When the secondary electron has a kinetic energy that is large enough to cause ionization and form its own "track," it is called a "delta ray" (δ-ray). This energy limit is about 100 eV.

Biological effects of radiation are generally considered dependent upon the deposition of energy in microscopic volumes generally estimated to be less than 1 μm in diameter (MT, Ch. 11). In tissue this distance corresponds to the range of a 6 keV electron. Hence, if a charged particle produces a δ-ray having an energy greater than about 10 keV, the energy cannot be considered "locally imparted." Customarily LET calculations have excluded energy associated with secondary particles above a given cutoff energy mΔ.* The δ-rays with energy in excess of mΔ are then treated as separate particles. The value for mΔ depends to a large extent on the size of the mass element being considered in the microscopic energy distribution. Figure 4.1 shows the variation in LET depending on the value chosen for mΔ for electrons and positrons. If mΔ is chosen equal to T'_{max}, the value obtained is called LET_∞ and is numerically equal to the stopping power. For heavy particles the difference between the stopping power (LET_∞) and LET_Δ is small for particle energies less than M (where M is the rest mass energy of the particle).

Since there is incomplete knowledge of the rate of energy loss of electrons having energies below a few keV (the process can no longer be treated as a collision between "free" electrons) only LET_∞ can be calculated to any significant degree of accuracy. In addition, the choice of the value Δ is rather arbitrary. Consequently

---

*The LET formula (Eq. 3-17) is in terms of $\tau = T/m$, thus Δ is in units of m.

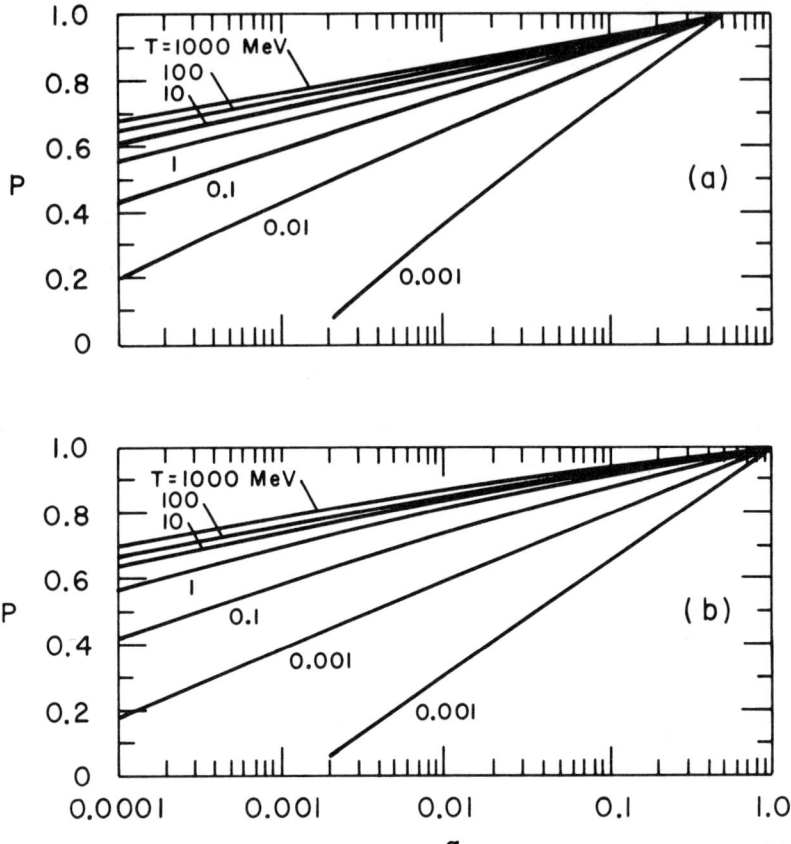

FIG. 4.1

The ratio $P = LET_\Delta/LET_\infty$ of electrons (a) and positrons (b) as a function of $g = m\Delta/T$. For electrons $g \leq 0.5$; for positrons $g \leq 1.0$. (Ref. 3.)

only LET_∞ has significant meaning in dosimetry applications. Therefore, for purposes of radiation protection the quality factor (QF) is related to LET_∞ only.

4.4 LET Distributions (ART, Ch. 2)

Linear energy transfer is the average rate of energy deposition by a particle of a particular energy. The application of LET to dosimetry is complicated by the fact that all charged particles traversing a mass element exposed to a particular radiation will not have the same energy (even if the incident radiation is mono-energetic). This energy spectrum of charged particles will lead to a LET distribution in the absorbing medium. A knowledge of the LET distribution can lead to an understanding of how the microscopic energy distribution varies with the incident radiation. The LET distributions can be expressed in several different ways.

One method is to define the fraction of particle track length T(L) at a given LET, L, per unit LET interval. Hence, T(L)dL expresses the relative amount of track in the LET interval between L and (L + dL).

A second method is to define the fraction of dose delivered D(L) at a given LET, L, per unit LET interval. Then, D(L)dL expresses the relative absorbed dose delivered in the LET interval between L and (L + dL).

Related to the second method is the definition of the energy dissipated N(L)L dL by electrons per unit volume in the LET interval between L and L + dL, where[1]

$$N(L) = -y(T)(dL/dT)^{-1} \quad (cm^{-2} - LET^{-1}) \quad (4.1)$$

The function y(T) is the electron fluence at energy T per energy interval resulting from the absorption of a given dose of X or γ radiation. The term $(dL/dT)^{-1}$ is derived from the formula for electron LET (Eq. (3.17) and (3.18)). The negative sign arises because dL/dT is negative since L is a decreasing function of energy.

It can be seen from the definitions that the three expressions are interrelated. Thus,

$$D(L) = \frac{N(L)L}{\int_{L_{min}}^{L_{max}} N(L)L \, dL} . \qquad (4.2)$$

Also, if one assumes that the total track length laid down in a volume of unit mass within the absorbing medium is K, then the length of track between L and L + dL is KT(L)dL. Multiplying this by L yields an energy representation and since we are considering a volume of unit mass, it also represents dose. Therefore,

$$L KT(L)dL = D(L)dL . \qquad (4.3)$$

This leads to a discussion of average values of LET. Since T(L) is a fractional track length,

$$\int_{L_{min}}^{L_{max}} LT(L)dL = \overline{L}_T \qquad (4.4)$$

the <u>track average LET</u>. Now from Eq. (4.3) we have;

$$K = \frac{\int_{L_{min}}^{L_{max}} D(L)dL}{\int_{L_{min}}^{L_{max}} LT(L)dL} = \frac{1}{\overline{L}_T} \qquad (4.5)$$

(Note: Since D(L) is defined as a fraction, $\int D(L)dL = 1$.) This leads to

$$D(L) = \frac{LT(L)}{\overline{L}_T} . \qquad (4.6)$$

We can also find the <u>dose average LET</u>

$$\overline{L}_D = \int_{L_{min}}^{L_{max}} LD(L)dL \qquad (4.7)$$

And the number average LET

$$\bar{L}_N = \frac{\int_{L_{min}}^{L_{max}} LN(L)dL}{\int_{L_{min}}^{L_{max}} N(L)dL} \tag{4.8}$$

These average values of LET can be used to determine such quantities as effective inactivation cross sections (σ_e) and effective RBE (R_e) under the assumption that σ_e and R_e are proportional to LET. For example, let us assume that RBE is a function of LET that can be expanded in a power series, i.e.,

$$r(L) = r_0 + r_1 L + r_2 L^2 + \ldots \tag{4.9}$$

and

$$R_e = \int_{L_{min}}^{L_{max}} D(L)\, r(L)dL \tag{4.10}$$

Then

$$R_e = \int_{L_{min}}^{L_{max}} r_0 D(L)dL + \int_{L_{min}}^{L_{max}} r_1 L D(L)dL + \int_{L_{min}}^{L_{max}} r_2 L^2 D(L)dL + \ldots$$

$$R_e = r_0 + r_1 \bar{L}_D + r_2 \bar{\bar{L}}_D + \ldots \tag{4.11}$$

where $\bar{L}_D$ and $\bar{\bar{L}}_D$ are the first and second moments of $D(L)$.

4.5 Event Size

At this point, it is useful to briefly discuss the concept of event size Y defined by Rossi[2] as the energy E_y deposited in a spherical volume of diameter d divided by d; that is,

$$Y = E_y/d \tag{4.12}$$

In the idealized case of straight particle tracks having uniform LET, Y has a constant value in spheres of different sizes. Actually, because of track curvature and δ-ray production, Y is generally not constant. This variation in Y expresses the general inadequacy of specifying local energy deposition in terms of LET_∞. The event size Y has a certain relationship to the LET and to the local energy density which will be discussed in Section 4.6.

If we consider a path length x through a spherical volume of diameter d traversed by particles of uniform LET, L, then the energy deposited in the volume

$$E_y = Lx$$

and

$$Y = L(x/d) . \qquad (4.13)$$

Hence,

$$Y_{max} = L \qquad (4.14)$$

4.6 Local Energy Density (Z) (ART)

Local energy density is the quantity of interest in the discussion of radiation effects on an absorbing medium. It is directly related to LET and event size Y as defined in Section 4.5. It also has a special relationship to absorbed dose.

If we consider an absorber of unit density material, an increment ΔZ of local energy density (ergs/g) is related to Y(keV/μm) and d(μm). When a single event of size Y occurs in a sphere of diameter d, the energy deposited is Yd in a volume equal to $(1/6)\pi d^3$. Thus,

$$\Delta Z = \frac{Yd}{\frac{1}{6}\pi d^3} \left(\frac{1.6 \times 10^{-9} \text{ erg/keV}}{10^{-12} \text{ cm}^3/\mu\text{m}^3} \right) \times \left(\frac{1}{1 \text{ g/cm}^3} \right)$$

or

$$\Delta Z = 3060 \, (Y/d^2) \, (\text{erg/g}). \qquad (4.15)$$

It should be noted that if the radiation is of high LET and d is small, ΔZ will represent a very appreciable local energy concentration. Figure 4.2 shows the maximum local energy density, ΔZ, in a 1μm sphere of tissue traversed by electrons or protons of various energies assuming $L = LET_\infty$. These curves have been calculated from Eq. (4.14) and (4.15) which give $\Delta Z = 3060$ L for x = d = 1μm.

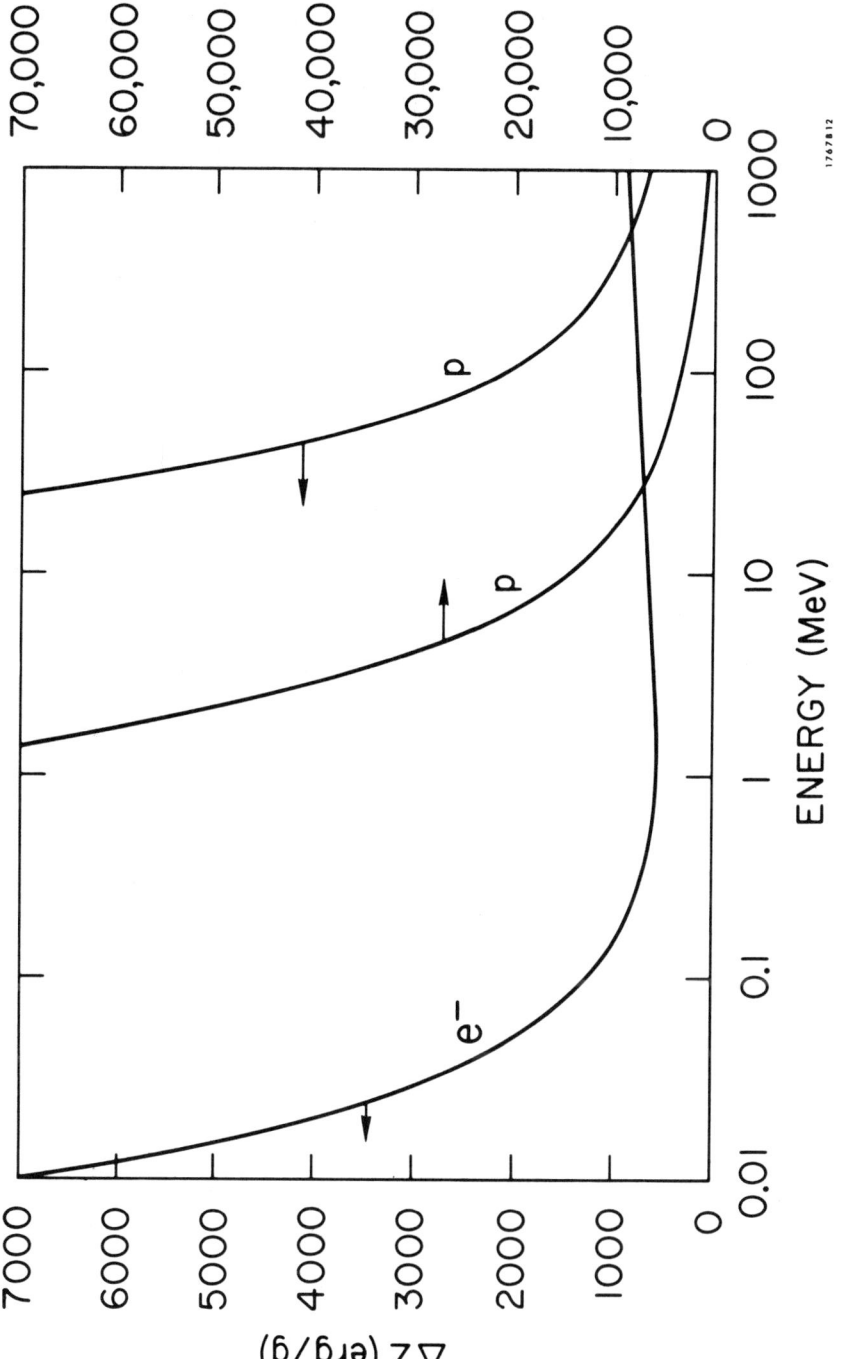

FIG. 4.2 Maximum energy density for protons and electrons traversing a 1 μm diameter sphere of tissue.

To compare the incremental local energy density rate for the traversal of a single particle with absorbed dose rate, consider the following situation. Particles (e^- or p) with energy of 1 MeV are incident on tissue. The absorbed dose rate is R = 1 Rad/hr. We consider a tissue sphere of d = 1 μm and let mΔ = 6 keV. Now for the electron, LET_6 = 0.14 keV/μm (using Fig. 4.1 and LET_∞ = 0.19 keV/μm).[3] For the proton $LET_6 = LET_\infty$ = 30 keV/μm[4]. Using these numbers we can calculate the maximum ΔZ from Eq. (4.15) and (4.14) and subsequently determine the rate of energy deposition. This is done in Table 4.3.

Table 4.3

Comparison of Energy Deposition Rate with Absorbed Dose Rate

Particle	R (erg/g-sec)	ΔZ (erg/g)	Particle Vel. (cm/sec)	Time to Traverse 1 μm (sec)	$\Delta Z/t$ (erg/g-sec)
e^-	2.8×10^{-2}	430	2.85×10^{10}	3.5×10^{-15}	1.2×10^{17}
p	2.8×10^{-2}	9.2×10^4	1.38×10^9	7.2×10^{-14}	1.3×10^{18}

Note: R = 1 rad/hour, Particle Energy = 1 MeV.

Since ΔZ represents the energy deposition from a single particle, it is obvious that the local energy deposition can be much larger than the macroscopic absorbed dose and the local energy deposition rate can be astronomical compared with the absorbed dose rate.

The energy density distribution is used to determine the frequency with which any event resulting in a particular ΔZ will occur. Now, we can define the fraction of dose delivered at an energy density ΔZ as $D(\Delta Z)$ so that for each rad of absorbed dose 100$D(\Delta Z)$ erg/g are delivered at an energy density ΔZ per unit ΔZ interval. Thus, the frequency of occurrence of events resulting in an energy density ΔZ per rad per unit ΔZ is

$$f(\Delta Z) = \frac{100\, D(\Delta Z)}{\Delta Z} \tag{4.16}$$

We can also express the frequency of energy increments ΔZ in terms of a probability of occurrence of an increment ΔZ, $P(\Delta Z)$ where:

$$D(\Delta Z) = \frac{P(\Delta Z)\Delta Z}{\int_0^{\Delta Z_{max}} P(\Delta Z)\, \Delta Z\, d(\Delta Z)} \tag{4.17}$$

So,

$$f(\Delta Z) = 100\, \frac{P(\Delta Z)}{\int_0^{\Delta Z_{max}} P(\Delta Z)\Delta Z\, d(\Delta Z)} \tag{4.18}$$

The frequency of events of all sizes is given by integration of $f(\Delta Z)$:

$$F = \int_0^{\Delta Z_{max}} f(\Delta Z)\, d(\Delta Z) = \frac{100}{\int_0^{\Delta Z_{max}} P(\Delta Z)\Delta Z\, d(\Delta Z)} \tag{4.19}$$

$$F = \frac{100}{\overline{\Delta Z}} \quad \text{(for 1 rad of absorbed dose)}$$

The quantity $P(\Delta Z)$ is the probability that an individual energy loss ΔZ will occur. The probability of finding an energy density Z at a specific point is denoted by $P(Z)$ which also is the relative frequency with which Z will be found in a large number of randomly selected spherical volumes in the irradiated medium. When the absorbed dose D is very low, the value of Z is much less than the mean value of ΔZ. The spherical volumes under consideration will be traversed only once or not at all by a charged particle. Thus, Z is due to single events and $P(Z)$ is proportional to $P(\Delta Z)$. On the other hand, when D is large, Z is caused by many individual traversals each of which deposit an increment ΔZ.

The distribution $P(Z)$ is an extremely complex function of the three variables Z, d, and D and as such does not have a simple analytical expression. It can be measured in certain cases by simulating the small (~1 μm) diameter tissue sphere with a spherical volume of tissue equivalent gas. The gas volume diameter and pressure are adjusted to properly simulate the unit density tissue sphere of interest. Thus, if the gas volume diameter is 10 cm and the unit density sphere to be simulated is 1 μm (10^{-4} cm), the gas density must be 10^{-5} g/cm^3.

Despite measurement and calculational difficulties some general features of the Z distributions can be discussed. First, we will consider the behavior of P(Z) as a function of d (the diameter of the sphere of interest). If d is large (or the order of millimeters), Z is always very nearly equal to D (i.e., the mass element is large enough to average out the individual variations in the locally deposited energy). Consequently, the curve of P(Z) as a function of Z will take the shape of a Gaussian distribution of narrow width about Z = D. As d is made smaller, the individual fluctuations in Z become more important and the width of the distribution will increase although the mean value of Z will remain equal to D. (This assumes that D is high enough to ensure a Gaussian distribution as discussed below.)

The behavior of P(Z) as a function of absorbed dose D shows that P(Z) is again Gaussian as long as D is large enough to ensure that the locally deposited energy Z is due to many events. This requires higher doses of high LET radiation than low LET radiation because the energy loss process is more uniform along the path of low LET radiations. The magnitude of the individual energy density increments ΔZ depends on the length of the charged particle path through the sphere of interest. From Eq. (4.15)

$$\Delta Z = 3060 \, (Y/d^2)$$

and since

$$Y = L(x/d)$$

from (4.13),

$$\Delta Z = 3060 \, L(x/d^3). \tag{4.20}$$

Hence, ΔZ also depends directly on the LET of the particle. Thus, although $P(Z)$ is Gaussian for large values of D, it becomes skewed as D becomes smaller and the individual increments ΔZ become more important. In fact, for low doses $P(Z)$ approaches $P(\Delta Z)$ which is highly skewed because of the high probability of ΔZ being zero and the fact that when an interaction does take place ΔZ will be very large compared with D. Some typical distributions are shown in Figs. 4.3 to 4.6.* The analytical details of these distributions were discussed in Sections 3.7 and 3.8. Since distributions in local energy density are intimately related to distributions in collision energy loss, they will be affected in the same way and exhibit either a Gaussian or Landau type of distribution.

4.7 Conclusions

When an absorbing medium is irradiated, the energy density is always non-uniform on a microscopic scale. Although the concepts, analytical treatment, and measurements involved are difficult, considerable progress has been made toward better definition and understanding of local energy density. More work needs to be done particularly in the area of applicability. It appears at this point that knowledge of the detailed distributions of LET and energy density might be most important in radiobiological research and radiation therapy. On the other hand, the formulation of present radiation protection recommendations and in particular, the definition of the dose equivalent make a measurement of the dose distribution in LET important whenever exposure to high-LET radiation occurs. These measurements are difficult at present primarily because of the complex and cumbersome equipment required.

---

*From (ART), Chapter 2.

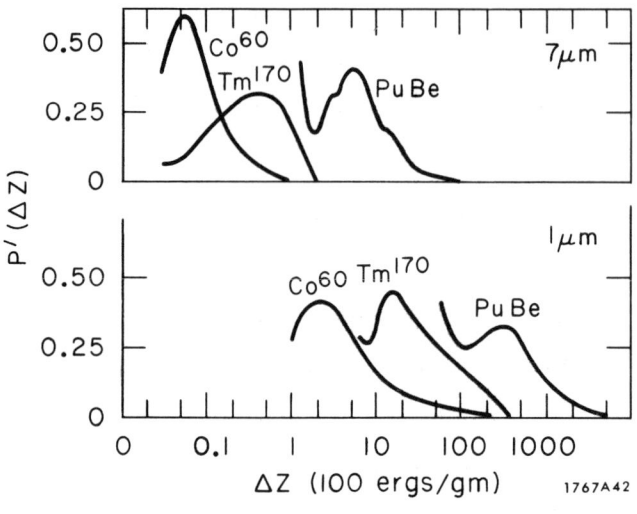

FIG. 4.3

Probability of increment ΔZ in 7 and 1 μm spheres.
$P'(\Delta A) = P(\Delta Z)/\Delta Z$

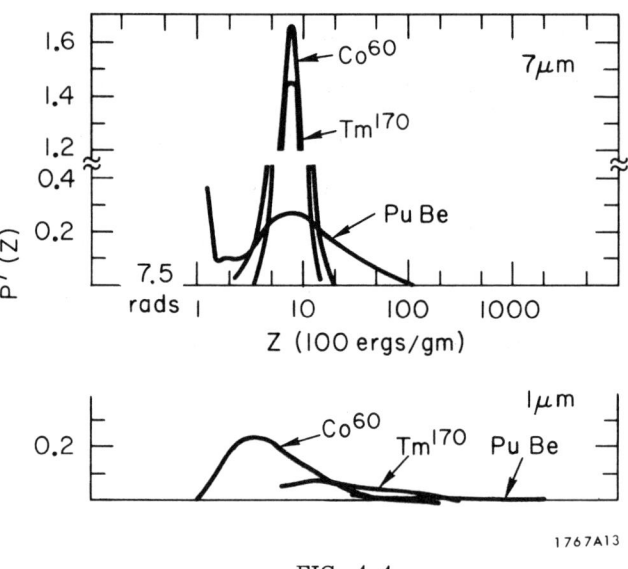

FIG. 4.4

Local energy density frequencies in 7 and 1 μm spheres. Dose = 7.5 rads.

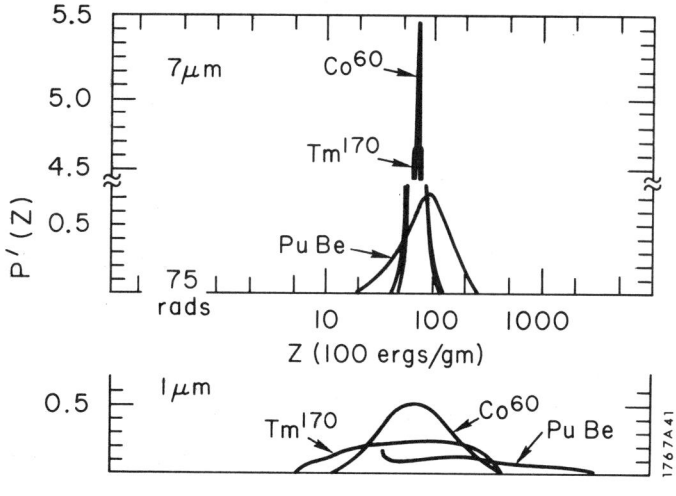

FIG. 4.5

Local energy density frequencies in 7 and 1 μm spheres. Dose = 75 rads.

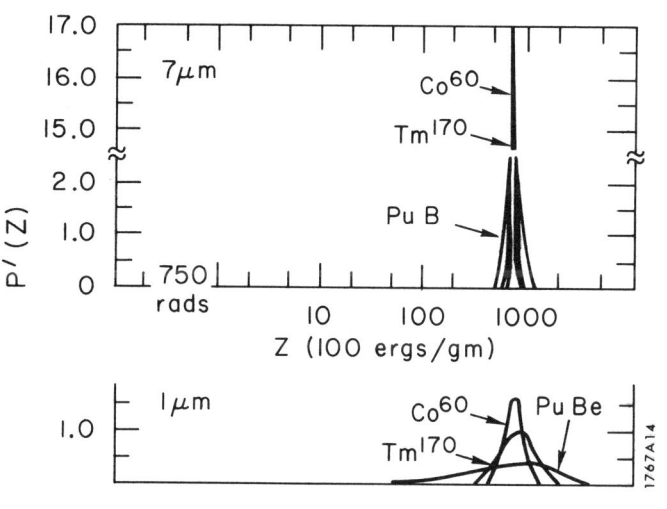

FIG. 4.6

Local energy density frequencies in 7 and 1 μm spheres. Dose = 750 rads.

REFERENCES

1. W. R. Bruce, M. L. Pearson and H. S. Freedhoff, Radiation Res. **19**, 606 (1963).
2. H. H. Rossi, M. H. Biavati and W. Gross, Radiation Res. **15**, 431 (1961).
3. M. J. Berger and S. M. Seltzer, <u>Tables of Energy Losses and Ranges of Electrons and Positrons,</u> National Aeronautics and Space Administration, Washington, D.C., (NASA-SP-3012) (1964).
4. W. H. Barkas and M. J. Berger, <u>Tables of Energy Losses and Ranges of Heavy Charged Particles,</u> National Aeronautics and Space Administration, Washington, D.C., (NASA-SP-3013) (1964).

MAIN REFERENCES

(MT) K. Z. Morgan and J. E. Turner (eds.), <u>Principles of Radiation Protection</u> (Krieger Publishing Co., New York, 1973).

(ART) F. H. Attix, W. C. Roesch, and E. Tochilin (eds.), <u>Radiation Dosimetry</u>, Second Edition, Volume I, Fundamentals (Academic Press, New York, 1968).

CHAPTER 5

DOSE CALCULATIONS

5.1 Introduction

In the previous sections we have studied in detail the interactions of photons and electrons as they pass through an absorbing medium. In this section we will develop the basic formulas for the calculation of radiation flux density and absorbed dose rate from external gamma radiation sources of various geometries. A determination of the absorbed dose requires a knowledge of the dose rate which in turn requires a knowledge of the source energy, the flux density and the rate of absorption of the radiation per unit path length about the point of interest. The flux density depends on the radiation source. We shall discuss gamma ray sources first, then will develop formulas for calculating the particle flux density for various sources, and finally will present the calculation of absorbed dose rate and dose.

5.2 Sources

Radiation sources can be characterized by their strength and their geometry. In our discussion we will be concerned primarily with photons and four source geometries with the following source strengths:

1) Point source S (photons $\sec^{-1}$)
2) Line source S_L (photons $cm^{-1}\text{-}\sec^{-1}$)
3) Area source S_A (photons $cm^{-2}\text{-}\sec^{-1}$)
4) Volume source S_V (photons $cm^{-3}\text{-}\sec^{-1}$)

To determine energy flux density we must multiply the source strength by the photon energy. In general, the source will not be monoenergetic, and consequently the source strength will be a function of energy. In our treatment we will develop the formulas for particle flux density for various source geometries. Since this is a purely geometric treatment, the source energy does not affect the formulation. We will point out the quantities in the formulas which are energy dependent and which would require an integration (or sum over discrete energies) in the case of a source that is not monoenergetic.

5.3 Flux Density

The general pattern to be followed in the development of the flux density formulas will be to derive the expression for flux density at a point assuming a uniform source distribution and neglecting: (1) Attenuation, (2) buildup, and (3) self absorption. We will then extend each development to include nonuniform source distribution, attenuation, buildup, and self absorption as appropriate.

The distribution of activity in a radioactive source is generally considered uniform unless otherwise specified. One case where the source distribution is not uniform is a nuclear reactor core, another is a pipeline carrying a short-lived radioactive isotope. Therefore, we will introduce calculations involving certain nonuniform source distributions.

Attenuation, of course, becomes an important parameter when absorbers are introduced between the source and the point of interest. Attenuation is generally energy dependent and any terms in which the attenuation coefficient appears will have to be included in the integration over energy for sources which are not monoenergetic. Buildup also becomes important when absorbers are present between the source and the point of interest. A detailed discussion of buildup is included later in Section 5.12. Buildup is also generally a function of energy and must be treated as such for any sources that are not monoenergetic.

Self absorption may be important in the consideration of volume sources particularly when the source dimensions are of the same order of magnitude as the photon mean free path in the material. Since this is an absorption process, it too is energy dependent.

5.4 Point Isotropic Source

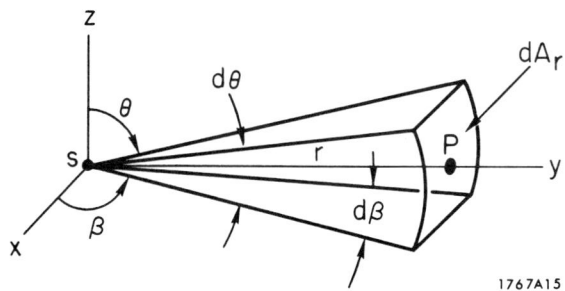

Although radiation sources have finite dimensions, they are often sufficiently small compared with the distance from the point of measurement to the source that the source may be treated as a point. The assumption of a point isotropic source implies that the source distribution is uniform and photon absorption in the source may be neglected. The photon flux density at point P is by definition the number of photons crossing the area A_r per unit time. Thus,

$$\phi = S/A_r$$

In spherical coordinates

$$A_r = \int_0^\pi \int_0^{2\pi} r \, d\beta \, r \sin\theta \, d\theta$$

$$= 4\pi r^2.$$

Thus

$$\phi = \frac{S}{4\pi r^2} \tag{5.1}$$

If an absorber is interposed between S and P we must account for the photon attenuation. Thus

$$\phi = \frac{S}{4\pi r^2} e^{-\mu t} \tag{5.2}$$

where μ is the appropriate mass attenuation coefficient and t is the absorber thickness in g/cm^2.

Buildup can be accounted for by simply multiplying Eq. (5.2) by the appropriate buildup factor B, so that

$$\phi = \frac{S}{4\pi r^2} B e^{-\mu t} . \qquad (5.3)$$

The various forms of B are discussed in Section 5.13.

5.5 Line Source

The formula for the flux density at a point P from a line source of length L depends on the location of P with respect to the line. Three points, as indicated in the following diagrams, will be considered.

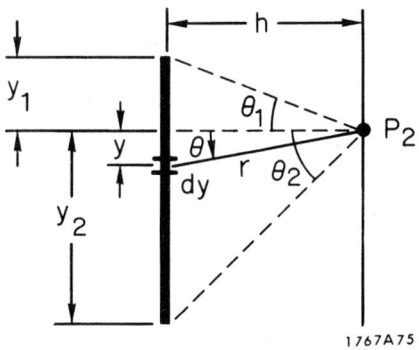

At P_2 the differential flux density from the line source element dy is given by

$$d\phi_2 = \frac{S_L \, dy}{4\pi r^2}$$

From the diagram it can be seen that

$$r = h \sec \theta$$
$$y = h \tan \theta$$
$$dy = h \sec^2 \theta \, d\theta$$

Thus

$$\phi_2 = \frac{S_L}{4\pi} \int_{-y_1}^{y_2} \frac{dy}{r^2} = \frac{S_L}{4\pi} \int_{-|\theta_1|}^{|\theta_2|} \frac{h \sec^2 \theta}{h^2 \sec^2 \theta} d\theta$$

$$= \frac{S_L}{4\pi h} (|\theta_2| + |\theta_1|) \tag{5.4}$$

Similarly for P_1

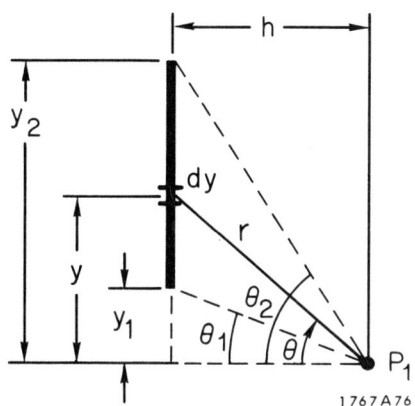

$$\phi_1 = \frac{S_L}{4\pi} \int_{y_1}^{y_2} \frac{dy}{r^2} = \frac{S_L}{4\pi h} (|\theta_2| - |\theta_1|) \tag{5.5}$$

For the situation where the point of interest P_3, is on the axis of the line source

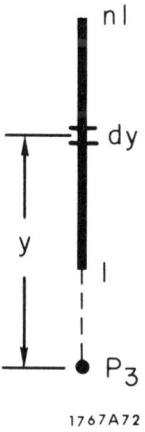

107

$$\phi_3 = \frac{S_L}{4\pi} \int_\ell^{n\ell} \frac{dy}{y^2}$$

$$= \frac{S_L}{4\pi} \left[-\frac{1}{y}\right]_\ell^{n\ell} = \frac{S_L}{4\pi} \left[\frac{1}{\ell} - \frac{1}{n\ell}\right]$$

$$= \frac{S_L}{4\pi n\ell} (n-1) \tag{5.6}$$

Now if we consider the situation in which absorbers have been added between the source and the point of interest we find that the integral in the case of P_2 becomes

$$\phi_2 = \frac{S_L}{4\pi h} \int_{-|\theta_1|}^{|\theta_2|} e^{-\sum_i \mu_i t_i \sec\theta} \, d\theta$$

where $t_i \sec\theta$ is the thickness of the ith absorber along the line from dy to P_2. This integral can be put into the form of the Sievert integral $F(\theta, \mu t)$ (see Appendix for graphs of $F(\theta, \mu t)$)

$$\phi_2 = \frac{S_L}{4\pi h} \left\{ \int_0^{|\theta_2|} e^{-\sum_i \mu_i t_i \sec\theta} \, d\theta + \int_0^{|\theta_1|} e^{-\sum_i \mu_i t_i \sec\theta} \, d\theta \right\}$$

$$= \frac{S_L}{4\pi h} \left\{ F(|\theta_2|, \sum_i \mu_i t_i) + F(|\theta_1|, \sum_i \mu_i t_i) \right\} \tag{5.7}$$

A similar equation can be derived for ϕ_1.

$$\phi_1 = \frac{S_L}{4\pi h} \left\{ F(|\theta_2|, \sum_i \mu_i t_i) - F(|\theta_1|, \sum_i \mu_i t_i) \right\} \tag{5.8}$$

Furthermore, Eq. (5.6) becomes

$$\phi_3 = \frac{S_L}{4\pi n\ell} (n-1) e^{-\sum_i \mu_i t_i} \tag{5.9}$$

Now, if we further include buildup, the form of the flux density equation will depend on the nature of the form chosen to represent the buildup. For example, if we choose a linear representation for buildup

$$B_1 \cong 1 + \alpha(E) \mu r$$

where

$$r = t \sec \theta \quad (t \leq h)$$

then, for a single absorber

$$\phi_2 = \frac{S_L}{4\pi h} \int_{-|\theta_1|}^{|\theta_2|} (1 + \alpha\mu t \sec \theta) e^{-\mu t \sec \theta} d\theta$$

$$\phi_2 = \frac{S_L}{4\pi h} \left\{ F(|\theta_2|, \mu t) + F(|\theta_1|, \mu t) + \alpha\mu t \int_{-|\theta_1|}^{|\theta_2|} \sec \theta \, e^{\mu t \sec \theta} d\theta \right\} \quad (5.10)$$

The integral in general must be evaluated numerically. Similar equations can be written for points P_1 and P_3.

We can now consider the case of a nonuniform source distribution. For example, consider the situation of a pipeline transporting a liquid containing a short-lived radioisotope and the point of observation is P_2. (See previous figure.) In this case,

$$S_L(y) = S_0 e^{-K(y_2 - y)} \quad \text{where}$$

$K = \lambda/v$

v = flow velocity (in negative y-direction on diagram)

$S_0 = A_0 \lambda/v$

A_0 = activity in the pipe at $y = y_2$.

If we assume that there is no attenuation or buildup, then

$$\phi_2 = \int_{-y_1}^{y_2} \frac{S_L(y)}{4\pi r^2} dy = \frac{A_0 \lambda e^{-Ky_2}}{4\pi h v} \int_{-|\theta_1|}^{|\theta_2|} e^{Kh \tan \theta} d\theta \quad (5.11)$$

which can be integrated numerically.

We can consider self absorption in the source in the case of P_3. This would introduce a term $e^{-\mu_s(y - \ell)}$ in the differential flux density so that

$$\phi_3 = \frac{S_L}{4\pi} \int_{\ell}^{n\ell} \frac{e^{-\mu_s(y - \ell)}}{y^2} dy$$

where μ_s is the linear attenuation coefficient of the source material. By a change of variables

$$\phi_3 = \frac{S_L}{4\pi} \mu_s \, e^{\mu_s \ell} \int_{\mu_s \ell}^{\mu_s n\ell} \frac{e^{-x}}{x^2} \, dx$$

$$= \frac{S_L}{4\pi} \mu_s \, e^{\mu_s \ell} \left\{ \int_{\mu_s \ell}^{\infty} \frac{e^{-x}}{x^2} \, dx - \int_{\mu_s n\ell}^{\infty} \frac{e^{-x}}{x^2} \, dx \right\}$$

$$\phi_3 = \frac{S_L}{4\pi} e^{\mu_s \ell} \left\{ \frac{1}{\ell} E_2(\mu_s \ell) - \frac{1}{n\ell} E_2(\mu_s n\ell) \right\} \tag{5.12}$$

where E_2 is an exponential integral (see Appendix for graphs of E_n).

The question often arises as to when a line source can be approximated by a point source. In the simplest geometrical situation we have (from Eq. (5.4)),

$$\phi_2 = \frac{S_L}{4\pi h} (|\theta_2| + |\theta_1|)$$

For small angles the approximation $\tan |\theta| = |\theta|$ can be made. In the case where $|\theta_1| = |\theta_2|$, $\tan |\theta_1| = \tan |\theta_2| = L/2h$; and so, $|\theta_1| + |\theta_2| = L/h$ in the small angle approximation. Thus

$$\phi_2 = \frac{S_L L}{4\pi h^2}$$

which is the equation for a point source with $S = S_L L$.

The small angle approximation is good to about 10% for $\theta < 30°$. Since $\tan 30° = 0.58$, a line source can be treated as a point source for values of $L/h < 1.2$ (or, when the separation distance h is greater than the length of the line L).

5.6 Area Source

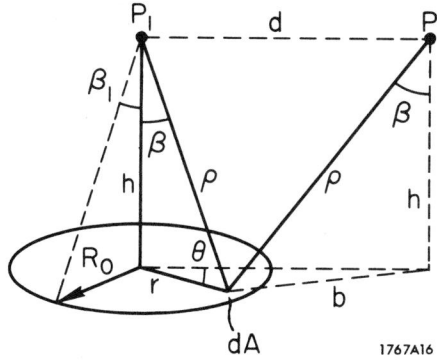

We now consider a source uniformly distributed over a plane as shown in the figure. The differential flux density at P is

$$d\phi = \frac{S_A \, dA}{4\pi \rho^2}$$

where,

$$dA = r\,dr\,d\theta$$
$$b^2 = d^2 + r^2 - 2rd\cos\theta$$
$$\rho^2 = h^2 + b^2 = h^2 + d^2 + r^2 - 2rd\cos\theta$$

Then the flux density at point P is given by:

$$\phi = \frac{S_A}{4\pi} \int_0^{R_0} \int_0^{2\pi} \frac{r\,dr\,d\theta}{h^2 + d^2 + r^2 - 2rd\cos\theta} \tag{5.13}$$

We can do the integration over θ by realizing the integral is of the form

$$\int_0^{2\pi} \frac{d\theta}{1 + a\cos\theta} = \frac{2\pi}{\sqrt{1 - a^2}}, \qquad (a^2 < 1)$$

Therefore

$$\phi = \frac{S_A}{2} \int_0^R \frac{r\,dr}{\left[(h^2 + d^2 + r^2)^2 - 4r^2 d^2\right]^{1/2}}$$

If we now let $x = r^2$, we have

$$\phi = \frac{S_A}{4} \int_0^{R_0^2} \frac{dx}{\sqrt{x^2 + 2(h^2 - d^2)x + (h^2 + d^2)^2}}$$

$$= \frac{S_A}{4} \ln\left|\frac{R_0^2 + h^2 - d^2 + \sqrt{(R_0^2 + h^2 - d^2)^2 + 4d^2 h^2}}{2h^2}\right| \tag{5.14}$$

In the special case of P_1 on the axis of the disc, $d = 0$. Hence,

$$\phi_1 = \frac{S_A}{4} \ln\left(\frac{R_0^2 + h^2}{h^2}\right) \tag{5.15}$$

If there is absorbing material between the source and the point of measurement, this can be taken into account as before by including a term $e^{-\sum_i \mu_i t_i \sec\beta}$ in the differential flux density equation. Now $\sec\beta = \rho/h$, so that

$$d\phi = \frac{S_A}{4\pi\rho^2} e^{-(\rho/h)\sum_i \mu_i t_i}\,dA$$

The integral of this equation is quite complicated and will not be discussed here. It is solved and the flux density equation given in Rockwell[1] (p. 394) for the two cases of

$$d^2 < R_0^2 + h^2 \quad \text{and} \quad d^2 > R_0^2 - h^2.$$

If we treat the case of P_1 on the axis of the disc, the equation simplifies to:

$$\phi_1 = \frac{S_A}{2} \int_0^{R_0} \frac{e^{-\sum_i \mu_i t_i \sec\beta}}{h^2 + r^2}\,r\,dr$$

and $\sec\beta = \rho/h$ as before. If we now substitute $y = (\sum_i \mu_i t_i \rho)/h$, the integral becomes

$$\phi_1 = \frac{S_A}{2} \int_{\sum_i \mu_i t_i}^{\sum_i \mu_i t_i \sec\beta_1} \frac{e^{-y}}{y}\,dy = \frac{S_A}{2} \left\{\int_{\sum_i \mu_i t_i}^{\infty} \frac{e^{-y}\,dy}{y} - \int_{\sum_i \mu_i t_i \sec\beta_1}^{\infty} \frac{e^{-y}\,dy}{y}\right\}$$

which is just the difference of two exponential integrals (see appendix for graphs of E_n).

So that

$$\phi_1 = \frac{S_A}{2}\left[E_1(\Sigma_i \mu_i t_i) - E_1(\Sigma_i \mu_i t_i \sec\beta_1)\right] \quad (5.16)$$

For an infinite plane source, $\sec\beta_1 \to \infty$, so that

$$\phi_1 \to \frac{S_A}{2} E_1(\Sigma_i \mu_i t_i).$$

At this point, we can account for buildup (or scattering) in each of the absorbers. Using the exponential approximation for buildup (MT)

$$B \simeq A e^{-\alpha_1 \mu t \sec\beta} + (1-A) e^{\alpha_2 \mu t \sec\beta},$$

we arrive at

$$\phi_1 = \frac{S_A}{2}\left\{A\left[E_1[(1+\alpha_1)\mu t] - E_1[(1+\alpha_1)\mu t \sec\beta_1]\right]\right.$$
$$\left. + (1-A)\left[E_1(1+\alpha_2)\mu t] - E_1[(1+\alpha_2)\mu t \sec\beta_1]\right]\right\} \quad (5.17)$$

Here we have simplified the expression by considering only a single absorber of thickness t, and we have taken the point of observation at P_1, on the axis of the disc.

We can also consider the linear representation of buildup

$$B \simeq 1 + \alpha \mu t \sec\beta$$

in which case the flux density at P_1 is

$$\phi_1 = \frac{S_A}{2}\left[E_1(\mu t) - E_1(\mu t \sec\beta_1) + \alpha(e^{-\mu t} - e^{-\mu t \sec\beta_1})\right] \quad (5.18)$$

In this case it is obvious that the scattered radiation simply adds the exponential terms into the equation for flux density. In the previous expression the contribution from scatter is included in the exponential integrals and is not explicitly isolated.

The treatment of flux density from a disc source can be extended to include the case of a nonuniform source distribution. We will consider a two dimensional Gaussian radial distribution ($\sigma_x = \sigma_y = \sigma$, $r = (x^2 + y^2)^{1/2}$). For simplicity we will consider only the flux density at P_1 a distance h above the position of maximum source strength. We will take the source to be infinite in extent and assume no attenuation. Then

$$\phi_1 = \frac{S_A}{4\pi} \int_0^{2\pi} d\theta \int_0^\infty \frac{e^{-r^2/\sigma^2}}{r^2 + h^2} r\, dr$$

$$= \frac{S_A}{2} \int_0^\infty \frac{e^{-r^2/\sigma^2}}{r^2 + h^2} r\, dr \tag{5.19}$$

We can reduce this integral to one of known form by two substitutions. Let

$$r^2 = \sigma^2 x$$

$$dr = \frac{\sigma}{2\sqrt{x}} dx$$

which gives

$$\phi_1 = \frac{S_A}{4} \int_0^\infty \frac{e^{-x}\, dx}{x + (h^2/\sigma^2)}$$

Now let

$$u = x + h^2/\sigma^2$$

$$du = dx$$

So that

$$\phi_1 = \frac{S_A}{4} e^{h^2/\sigma^2} \int_{h^2/\sigma^2}^\infty \frac{e^{-u}}{u} du = \frac{S_A}{4} e^{h^2/\sigma^2} E_1(h^2/\sigma^2) \tag{5.20}$$

There is some point from the disc beyond which the disc may be treated as a point source $S = S_A \pi R_0^2$. The point source equation is then

$$\phi = \frac{S_A \pi R_0^2}{4\pi h^2} = \frac{S_A}{4} \frac{R_0^2}{h^2}$$

If we choose the point P_1, we are interested in the distance h for which the approximation

$$\frac{S_A}{4} \frac{R_0^2}{h^2} \simeq \frac{S_A}{4} \ln\left(\frac{R_0^2 + h^2}{h^2}\right)$$

(from Eq. (5.15) holds. Thus

$$\frac{R_0^2}{h^2} \approx \ln\left(\frac{R_0^2 + h^2}{h^2}\right)$$

or

$$-\frac{R_0^2}{h^2} \approx 1 - e^{R_0^2/h^2}$$

This approximation is good to within 10% when $R_0^2/h^2 < 0.20$ (or $R_0/h < 0.45$). Thus a disc source may be treated as a point source for $h > 2.2\, R_0$ (or when the separation distance h is greater than the source diameter $2R_0$).

5.7 Infinite Slab Source

The logical extension of the infinite area source is the infinite slab source. We now consider an area of infinite extent but of a finite thickness. In this case self absorption in the source material must be considered from the beginning. We can derive the flux density by a simple extension of the area source by writing the differential flux density from an infinite area element located at a distance x within the slab.

From Eq. (5.16) (extended to an infinite plane)

$$d\phi = \frac{S_V}{2} E_1\left[\mu t + \mu_s(h - x)\right] dx$$

Here $S_V dx = S_A$, h is now the slab thickness, and μ_s is the linear attenuation coefficient of the source material. We also consider attenuation in a single absorber positioned between the slab and the observation point (but we neglect buildup). By integrating over x, we have

$$\phi = \frac{S_V}{2} \int_0^h E_1\left[\mu t + \mu_s(h - x)\right] dx.$$

Substituting $y = \mu t + \mu_s(h - x)$ we get

$$\phi = \frac{S_V}{2\mu_s} \int_{\mu t}^{\mu t + \mu_s h} E_1(y) dy.$$

Realizing that

$$\int_{\mu t}^{\infty} E_1(y)dy = E_2(\mu t)$$

this becomes

$$\phi = \frac{S_V}{2\mu_s} \left[E_2(\mu t) - E_2(\mu t + \mu_s h) \right] \qquad (5.21)$$

We have assumed also that S_V is constant through the slab.

It is instructive to derive this result from the basic geometry.

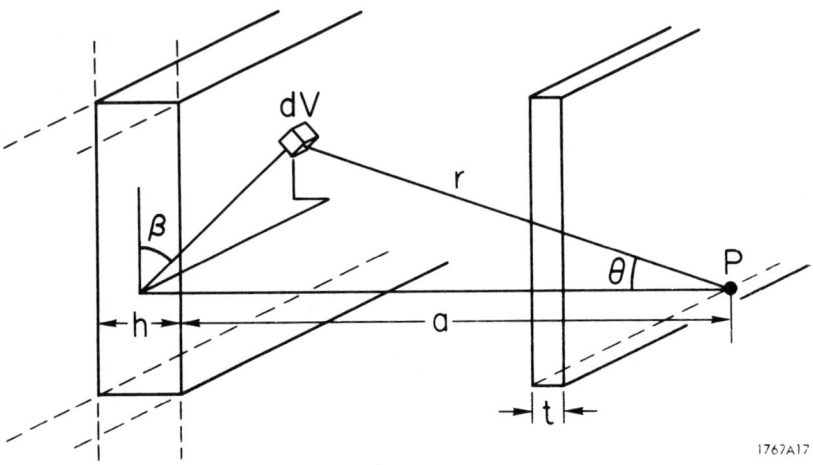

From the figure we see that the differential flux density at P is

$$d\phi = \frac{S_V}{4\pi r^2} dV \, e^{-(r-a\sec\theta)\mu_s} e^{-\mu t \sec\theta}$$

In spherical coordinates

$$dV = r^2 \sin\theta \, d\theta d\beta dr$$

So the flux density is given by

$$\phi = \frac{S_V}{4\pi} \int_0^{2\pi} d\beta \int_0^{\pi/2} \sin\theta \, d\theta \int_{a\sec\theta}^{(a+h)\sec\theta} \frac{r^2 e^{-[(r-a\sec\theta)\mu_s + \mu t \sec\theta]}}{r^2} dr$$

$$= -\frac{S_V}{2} \int_0^{\pi/2} \frac{\sin\theta \, d\theta}{\mu_s} e^{\mu_s a \sec\theta - \mu t \sec\theta} \left[e^{-\mu_s r} \right]_{a\sec\theta}^{(a+h)\sec\theta}$$

$$= \frac{S_V}{2\mu_s} \int_0^{\pi/2} \sin\theta \left[e^{-\mu t \sec\theta} - e^{-(\mu t + \mu_s h)\sec\theta} \right] d\theta$$

Now if we substitute $y = \mu t \sec\theta$ in the first integral and $y = (\mu t + \mu_s h)\sec\theta$ in the second integral we obtain

$$\phi = \frac{S_V}{2\mu_s} \left\{ \mu t \int_{\mu t}^{\infty} \frac{e^{-y}}{y^2} dy - (\mu t + \mu_s h) \int_{\mu t + \mu_s h}^{\infty} \frac{e^{-y}}{y^2} dy \right\}$$

So

$$\phi = \frac{S_V}{2\mu_s} \left[E_2(\mu t) - E_2(\mu t + \mu_s h) \right] \qquad (5.22)$$

which is identical to Eq. (5.21).

The properties of $E_2(x)$ are such that as $x \to \infty$, $E_2(x) \to 0$ and as $x \to 0$, $E_2(x) \to 1$, so that for a slab source of <u>infinite thickness</u> ($h \to \infty$), $E_2(\mu t + \mu_s h) \to 0$ and

$$\phi = \frac{S_V}{2\mu_s} E_2(\mu t) \qquad (5.23)$$

Similarly if there is <u>no absorber</u> between the slab source and P, then $\mu t = 0$ and therefore

$$\phi = \frac{S_V}{2\mu_s} \left[1 - E_2(\mu_s h) \right] \qquad (5.24)$$

From this it follows that the primary photon flux density at the surface of an infinitely thick slab source of infinite extent is

$$\phi = \frac{S_V}{2\mu_s} \qquad (5.25)$$

(that is, $\mu t = 0$ and $\mu_s h \to \infty$).

Up to this point we have not considered buildup. This can be taken into account by inserting any of the accepted representations for buildup into the differential flux density equation and carrying out the integration. This becomes quite complicated and we will simply write the result assuming the linear buildup representation (MT)

$$B \simeq 1 + \alpha_1 \mu_s (r - a\sec\theta) + \alpha_2 \mu t \sec\theta$$

Namely,

$$\phi = \frac{S_V}{2\mu_s} \left\{ (1 + \alpha_1) \left[E_2(\mu t) - E_2(\mu t + \mu_s h) \right] \right.$$

$$\left. + \alpha_2 \mu t\, E_1(\mu t) - (\alpha_2 \mu t + \alpha_1 \mu_s h)\, E_1(\mu t + \mu_s h) \right\} \qquad (5.26)$$

Expressions can also be derived assuming that the source varies with distance in the slab. Expressions for different types of source variations are given in Rockwell[1] and will not be discussed here.

5.8 <u>Right-Circular Cylinder Source: Infinite-Slab Shield, Uniform Activity Distribution</u>

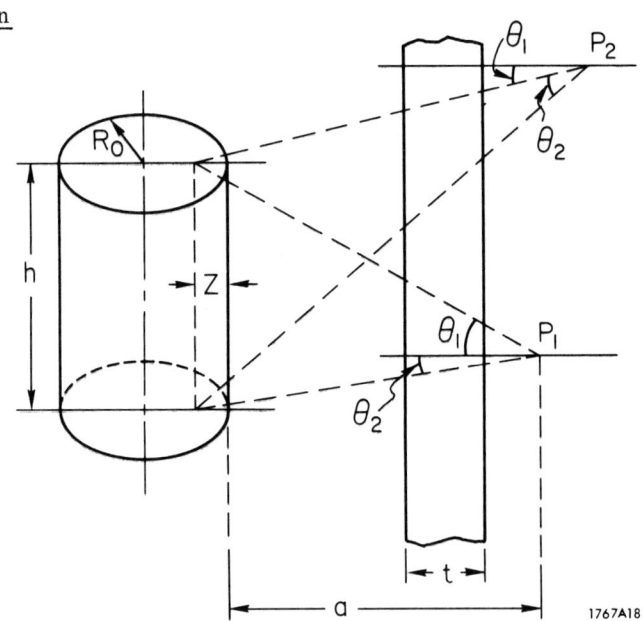

The slab absorber is parallel to the cylinder axis. The source strength per unit volume, S_V, is constant. The exact solution[3] of this problem is very lengthy and is not generally used. What is usually done[1] (FBM) is to approximate the cylinder by a line source of strength $S_L = \pi R_0^2 S_V$ which is positioned within the cylinder to correctly account for self absorption. There is no simple expression for $Z = Z(R_0, a, b)$, the self absorption distance; however, by empirically fitting the approximate method to the exact calculations,[3] only three curves for Z plus the $F(\theta, b)$ curves (that is, the Sievert integrals) for line sources are needed in order to solve cylinder-slab problems. The three curves needed to obtain Z are given in the Appendix and are used as follows:

CASE: $a/R_0 \geq 10$

Use figure A.20 (see Appendix) and $\mu_s R_0$ to obtain $\mu_s Z$, where $\mu_s (cm^{-1})$ is the macroscopic source attenuation coefficient. Then obtain b_2 from

$$b_2 = b_1 + \mu_s Z \tag{5.27}$$

where

$$b_1 = \sum_i \mu_i t_i \tag{5.28}$$

Finally, obtain the flux density at P_1 from

$$\phi_1 = \frac{S_V R_0^2}{4(a+Z)} [F(\theta_1, b_2) + F(\theta_2, b_2)] \tag{5.29}$$

and at P_2 from

$$\phi_2 = \frac{S_V R_0^2}{4(a+Z)} [F(\theta_2, b_2) - F(\theta_1, b_2)] \tag{5.30}$$

using the F-functions which are plotted in the Appendix. These estimates of the flux density are supposedly good to ± 10%,* provided $a/R_0 \geq 10$.

---

*Note: Provided that the correct buildup factors have been included.

CASE: $a/R_0 < 10$

Use Figs. A.21 and A.22 in conjunction with each other to obtain $\mu_s Z$. That is, knowing R_0, a, and μ_s, find m from the first graph; knowing a/R_0 and b_1, find $\mu_s Z/m$ from the second graph; then multiply these together to obtain $\mu_s Z$. Finally, follow the recipe above to obtain ϕ. This approximation will be good to +40% and −5%.

Other formulas are given for cylinders viewed exterior on end, and interior[1] (FBM).

Example:

Consider a cylindrical tank containing radioactive water uniformly distributed throughout. The field positron is P_1 with $\theta_1 = \theta_2$, and the distance is restricted to

$$R_0 \leq a \leq 70.0 \text{ inches}$$

with

1) $R_0 = 5.5$ inches
2) $h = 14.0$ inches
3) no shielding or buildup
4) self absorption in the water

and

5) the radioactive source consists mainly of 0.511 MeV photons with $\mu_s = 0.092 \text{ cm}^{-1}$ (the total attenuation coefficient for water).

The normalized flux density is obtained from Eq. (5.29), and is

$$2\phi/S_V R_0^2 = \frac{F(\theta, b_2)}{a + Z}$$

where we have dropped the subscript on theta, and where

$$\tan \theta = h/2(a + Z)$$

$$b_2 = \mu_s Z .$$

Using Figs. A.14, 20, 21, and 22, we obtain Table 5.1.

Table 5.1

a(in)	a(cm)	a/R_0	$\mu_s(a+R_0)$	m	$(1/m)\mu_s Z$	$\mu_s Z$	Z(cm)	$\tan\theta$	$\theta°$	$F(\theta, b_2)$	$2\phi/(S_V R_0^2)$
5.5	14.0	1.00	2.58	0.58	1.47	0.85	9.2	0.767	39.4	3.6×10^{-1}	1.6×10^{-2}
10.0	25.4	1.82	3.62	0.71	1.24	0.88	9.6	0.507	26.8	1.85×10^{-1}	5.3×10^{-3}
20.0	50.8	3.64	5.97	0.92	0.98	0.90	9.8	0.293	17.4	1.2×10^{-1}	2.0×10^{-3}
27.0	68.6	4.90	7.59	1.12	0.86	0.96	10.4	0.226	12.7	8.4×10^{-2}	1.07×10^{-3}
35.0	88.9	6.37	9.48	1.49	0.76	1.13	12.3	0.176	10.0	5.6×10^{-2}	5.5×10^{-4}
45.0	114.0	8.18	11.80	1.78	0.63	1.12	12.2	0.141	8.0	4.4×10^{-2}	3.4×10^{-4}
55.0	140.0	10.00	–	–	–	0.68	7.4	0.121	6.9	6.0×10^{-2}	4.1×10^{-4}
62.0	158.0	11.30	–	–	–	0.68	7.4	0.108	6.4	5.3×10^{-2}	3.2×10^{-4}
70.0	178.0	12.70	–	–	–	0.68	7.4	0.096	5.5	4.8×10^{-2}	2.6×10^{-4}

The data in Table 5.1 are plotted in Fig. 5.1 where they are compared with experimental data that was obtained as follows.

A tank, having the above dimensions, was filled with water taken from a SLAC* beam dump that had been operating for several hours with a high energy (E > 10 GeV) electron beam at a steady power level of 30 kW. The tank was returned to the laboratory and allowed to sit until the dominant activity $\left(^{15}\text{O}, T_{1/2} = 2 \text{ minutes}\right)$ had decayed away. The dominant activity was then ^{11}C which is a positron emitter (therefore, 0.511 MeV annihilation quanta) with a half life of 20 minutes. Measurements were quickly made (over a few minutes), as a function of distance from the tank, using a GM counter. The data are plotted in Fig. 5.1 (normalized at a = 20 inches). The comparison is reasonably good considering that buildup was excluded from the calculation and the GM counter probably doesn't correctly measure the photon flux density.

5.9 Spherical Source: Infinite-Slab Shield, Uniform Activity Distribution

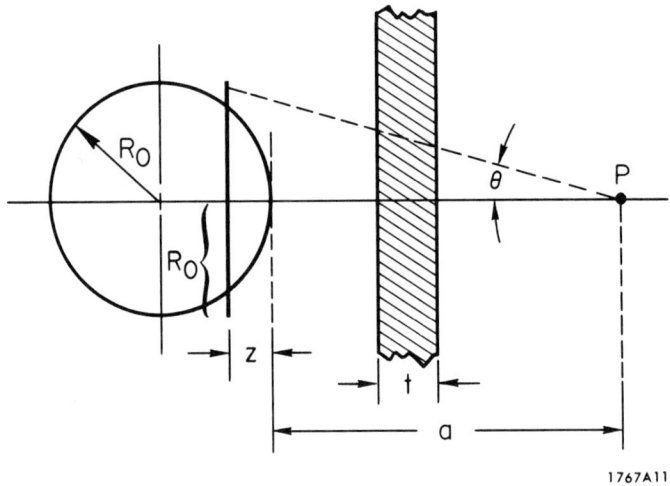

---
*Stanford Linear Accelerator Center.

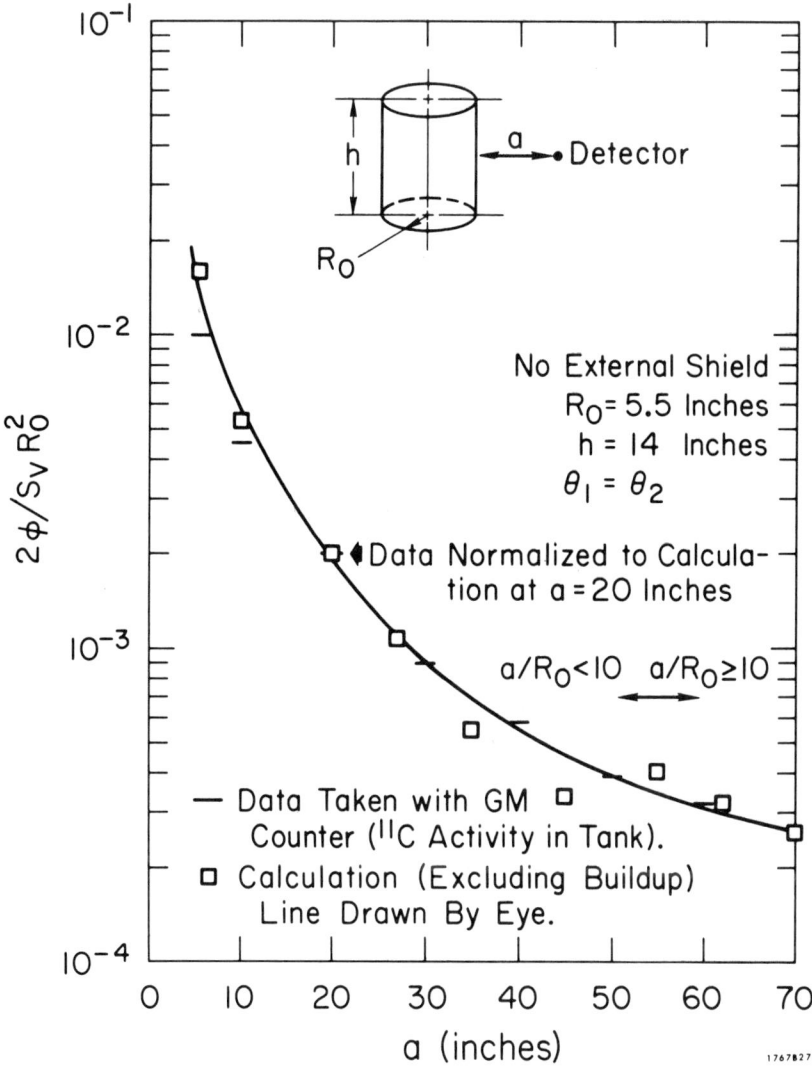

FIG. 5.1

Radiation at various distances from a right circular cylindrical source.

The following discussion holds only for a shield perpendicular to the radial vector from the sphere center to P. Again, as in the case of the cylinder above, the exact calculation is lengthy. The solution is approximated by replacing the sphere by an appropriately located disk source of radius R_0 (that is, the sphere radius), which has a source strength per unit area

$$S_A = 4 R_0 S_V/3$$

The self absorption distance, Z, is obtained by empirical fitting, using the exact calculations.[1] Figures A.23 and A.24 can be used to obtain Z as follows:

CASE: $a/R_0 < 1$

Use Fig. A.23 with $\mu_s(a + R_0)$ to obtain Z/R_0, and hence, Z. Then calculate b_2 from

$$b_2 = b_1 + \mu_s Z$$

where

$$b_1 = \sum_i \mu_i t_i .$$

Finally, calculate the flux density from

$$\phi = \frac{2}{3} S_V R_0 [E_1(b_2) - E_1(b_2 \sec \theta)] \qquad (5.31)$$

Range of Accuracy: -20% to +50%.

CASE: $a/R_0 \geq 1$

Use Fig. A.24 with $\mu_s R_0$ to obtain $\mu_s Z$, and then follow the above recipe to obtain ϕ. Range of accuracy: -5% to +15%.

Example:

Consider a sphere containing radioactive water uniformly distributed throughout and with no shield between source and detector. Take

$$R_0 = 7.0 \text{ inches}$$

$$3.5 \leq a \leq 70.0 \text{ inches}$$

$$\mu_s = 0.092 \text{ cm}^{-1} \text{ (0.511 MeV photons in water)}$$

The normalized flux density is obtained from Eq. (5.31) and is

$$3\phi/2S_V R_0 = E_1(b_2) - E_1(b_2 \sec \theta)$$

where

$$b_2 = \mu_s Z$$

and

$$\tan \theta = R_0/(a + Z)$$

Using Figs. A.23, A.24, and A.2, we obtain Table 5.2.

The data in Table 5.2 are plotted in Fig. 5.2 where they are compared with experimental data that was obtained in a manner similar to that described in Section 5.8, but using a hemispherical tank. The data were normalized to the calculation at a = 4 inches.

A point source, corresponding to $\phi \sim 1/a^2$, is plotted as the straight line in Fig. 5.2.

5.10 Spherical Source: Field Position at Center of Sphere

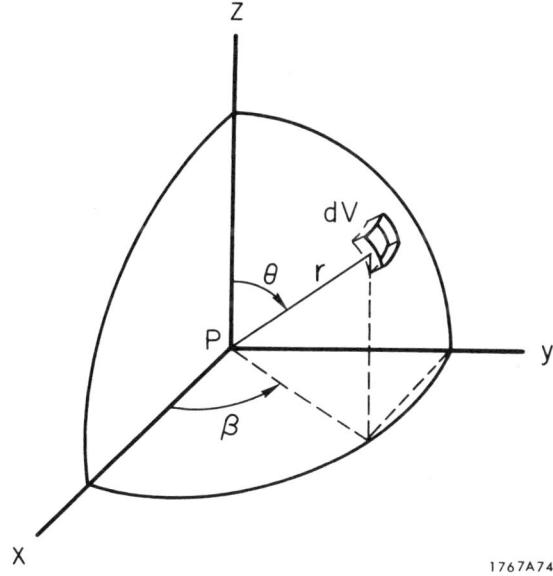

Table 5.2

a(in)	a(cm)	a/R_0	$\mu_s(a+R_0)$	Z/R_0	$b_2 = \mu_s Z$	Z(cm)	$\tan\theta$	$\theta(°)$	$b_2\sec\theta$	$E_1(b_2)$	$E_1(b_2\sec\theta)$	$3\phi/2S_V R_0$
3.5	8.89	0.5	2.46	0.598	0.98	10.6	0.912	42.4	1.33	0.22689	0.12933	9.8×10^{-2}
7.0	17.80	1.0	—	—	1.01	11.0	0.618	31.6	1.19	0.21574	0.16094	5.5×10^{-2}
14.0	35.50	2.0	—	—	1.01	11.0	0.383	20.9	1.08	0.21574	0.19216	2.4×10^{-2}
21.0	53.30	3.0	—	—	1.01	11.0	0.277	15.5	1.05	0.21574	0.20187	1.4×10^{-2}
28.0	71.00	4.0	—	—	1.01	11.0	0.217	12.3	1.03	0.21574	0.20867	7.1×10^{-3}
42.0	107.00	6.0	—	—	1.01	11.0	0.151	8.6	1.02	0.21574	0.21217	3.6×10^{-3}
56.0	142.00	8.0	—	—	1.01	11.0	0.116	6.6	1.0167	0.21574	0.21333	2.4×10^{-3}
70.0	178.00	10.0	—	—	1.01	11.0	0.0941	5.4	1.0145	0.21574	0.21412	1.6×10^{-3}

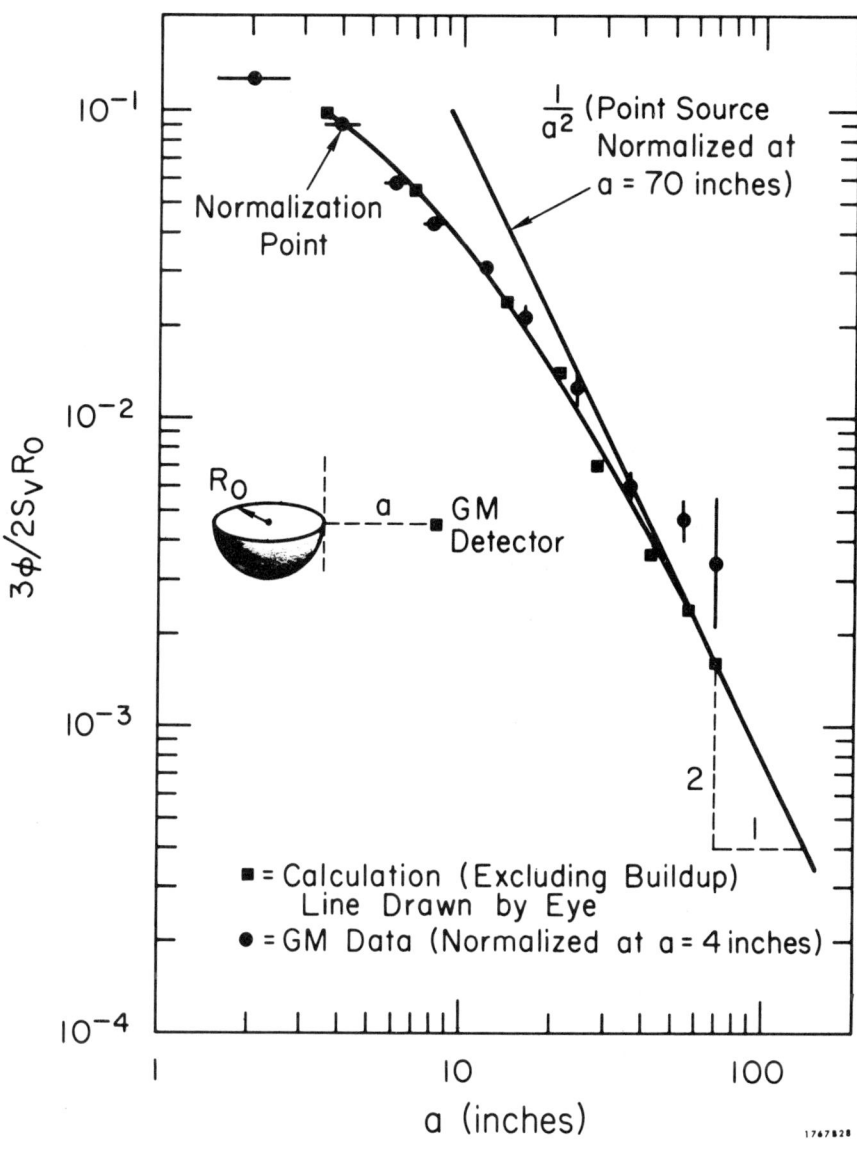

FIG. 5.2

Radiation at various distances from a hemispherical source.

$$d\phi = \frac{S_V e^{-\mu_s r} dV}{4\pi r^2} = \frac{S_V e^{-\mu_s r} r^2 \sin\theta \, dr \, d\theta \, d\beta}{4\pi r^2}$$

$$\phi = \frac{S_V}{4\pi} \int_0^{R_0} \int_0^{\pi} \int_0^{2\pi} e^{-\mu_s r} \sin\theta \, dr \, d\theta \, d\beta$$

$$= \frac{S_V}{\mu_s} \left(1 - e^{-\mu_s R_0}\right) \tag{5.32}$$

This submersion situation is applicable to finding the dose rate in a radioactive cloud or in a body of contaminated water. By symmetry, the uncollided flux from a hemisphere (that is, no buildup) is exactly one-half of this.

5.11 Transport of Radiation

The most thorough description of a radiation field (gamma, neutron, muon, etc.) consists in specifying at each point the number of particles per unit time going in each direction in each energy interval. We can define the <u>number flux density</u> by

$\phi(\bar{r}, E, \hat{\Omega})dEd\hat{\Omega}$ = number of particles at $\bar{r}$, with energy in dE about E and within the element of solid angle $d\hat{\Omega}$ about $\hat{\Omega}$ (unit vector direction), which cross in unit time a differential element of area whose normal is in the direction $\hat{\Omega}$.

We also can define an <u>energy flux density</u> by

$$I(\bar{r}, E, \hat{\Omega}) = E\phi(\bar{r}, E, \hat{\Omega}) \tag{5.33}$$

which gives the energy carried by particles rather than their number.

The equation that governs the transport of particles through matter, assuming that equilibrium in time has been established, is given by

$$\bar{\nabla} \cdot \hat{\Omega} \, \phi(\bar{r}, E, \hat{\Omega}) + \mu\phi(\bar{r}, E, \hat{\Omega})$$

$$= \iint \phi(\bar{r}, E', \hat{\Omega}')n\sigma(\hat{\Omega}' \to \hat{\Omega}, \, E' \to E)dE' \, d\hat{\Omega}' + s(\bar{r}, E, \hat{\Omega}) \tag{5.34}$$

where μ = attenuation (or interaction) coefficient

$s(\bar{r}, E, \hat{\Omega})$ = source number density

= number of particles created per unit time at $\bar{r}$ which move in the direction $\hat{\Omega}$ per unit solid angle with energy in dE about E.

n = number of scatterers per unit volume at r

= $ZN_0\rho/A$ electrons/cm^3 for photons

$\sigma(\hat{\Omega}' \to \hat{\Omega}, E' \to E)$ = differential cross section for scattering from $\hat{\Omega}'$ to $\hat{\Omega}$ and from E' to E.

Equation (5.34) is Boltzmann's integro-differential equation for the number flux density under time-equilibrium conditions. Multiplying through by E and using Eq. (5.33) we obtain

$$\bar{\nabla} \cdot \hat{\Omega} I(\bar{r}, E, \hat{\Omega}) + \mu I(\bar{r}, E, \hat{\Omega})$$

$$= \iint I(\bar{r}, E', \hat{\Omega}') n \sigma(\hat{\Omega}' \to \hat{\Omega}, E' \to E) \frac{E \, dE'}{E'} d\hat{\Omega}' + S(\bar{r}, E, \hat{\Omega}) \quad (5.35)$$

where S = Es which is the energy flux density form of the steady-state Boltzmann transport equation.

Now, for the case of photons, the differential scattering cross section is obtained from the Klein-Nishina formula and from the kinematic equations relating energy and angle for Compton scattering of photons from free electrons (see Section 2.4c). A modified form of Eq. (5.35) is the basis for all calculations in gamma-radiation dosimetry. Because of its basic complexity, however, the transport equation is never solved in closed form in practical cases. The greatest use of the transport equation arises in estimating buildup factors that are applied to the results of calculations based on the uncollided-flux approximation — such as the source geometry computations that were carried out in Sections 5.4 through 5.10.

It is the Compton interaction process that makes calculations (and sometimes experiments) so prohibitively difficult. To appreciate this fully, let us assume that

$\sigma(\hat{\Omega}' \to \hat{\Omega}, E' \to E)$ equals zero in Eq. (5.35). In other words, we will make the uncollided (or unscattered) flux approximation; that is, no Compton scattering back into the point detector, although Compton scattering out can be considered in the attenuation coefficient, if so desired. Equation (5.35) becomes

$$\nabla \cdot \hat{\Omega} I(\bar{r}, E, \hat{\Omega}) + \mu I(\bar{r}, E, \hat{\Omega}) = S(\bar{r}, E, \hat{\Omega}) \tag{5.36}$$

which is a linear inhomogeneous partial differential equation.

As an example of the solution of this equation, let us calculate the energy flux density, $I(r, E)$, for a monoenergetic isotropic point source located at $\bar{r} = 0$. In this case,

$$S(\bar{r}, E, \hat{\Omega}) = 0 \quad \text{for } |\bar{r}| > 0$$

$$\hat{\Omega} = \hat{r} \tag{5.37}$$

and

$$\nabla = \frac{\hat{r}}{r^2} \frac{\partial}{\partial r} r^2 + \frac{\hat{\theta}}{r \sin \theta} \frac{\partial}{\partial \theta} \sin \theta + \frac{\hat{\phi}}{r \sin \theta} \frac{\partial}{\partial \phi} \tag{5.38}$$

so that Eq. (5.36) reduces to

$$\frac{1}{r^2} \frac{d}{dr} (r^2 I) + \mu I = 0, \quad r > 0 \tag{5.39}$$

which has the solution

$$I(r, E) = \frac{C e^{-\mu(E) r}}{r^2} \quad (\text{MeV/cm}^2 - \text{sec}) \tag{5.40}$$

where C is a constant of integration. If we let $C = SE/4\pi$, we obtain:

$$\phi = \frac{S}{4\pi r^2} e^{-\mu r} \tag{5.41}$$

which is similar to Eq. (5.2) of Section 5.4. It should be pointed out that in the derivations that were made in Sections 5.4 through 5.10, we used the point-source formula (5.2) as a starting point rather than Eq. (5.36). In effect, we made a linear superposition of isotropic point sources, which was valid since Eq. (5.36) is a linear partial differential equation.

5.12 Buildup Factor Corrections to the Uncollided-Flux Density Calculations

Because the uncollided flux density approximation neglects photons that undergo Compton scattering-in interactions, such quantities as exposure rate, absorbed dose rate and flux density (number and energy) are underestimated in uncollided flux density calculations. The degree of underestimation depends on several factors, the most important being the energy of the primary photons and the distance (μr) from source-to-detector. In order to correct for Compton scattering into the detector, one introduces the concept of buildup. For each physical quantity of interest, one can define a buildup factor. Thus, we have:

Number Flux Density Buildup Factor

$$B(\bar{r}) = \frac{\int \phi(\bar{r}, E) dE}{\int \phi_0(\bar{r}, E) dE} \tag{5.42}$$

Energy Flux Density Buildup Factor

$$B_E(\bar{r}) = \frac{\int I(\bar{r}, E) dE}{\int I_0(\bar{r}, E) dE} \tag{5.43}$$

Absorbed Dose Buildup Factor

$$B_a(\bar{r}) = \frac{\int \frac{\mu_{en}(E)}{\rho} I(\bar{r}, E) dE}{\int \frac{\mu_{en}(E)}{\rho} I_0(\bar{r}, E) dE} \tag{5.44}$$

where the zero subscript indicates the uncollided flux density (number or energy). Generally, B_a is defined using μ_a/ρ instead of μ_{en}/ρ, where

$$\mu_{en}/\rho = \frac{1}{\rho}\left[\tau(1-f) + \frac{\sigma \bar{E}}{h\nu} + \kappa\left(1 - \frac{2mc^2}{h\nu}\right)\right] \times (1-G) \tag{5.45}$$

and

$$\mu_a/\rho = \frac{1}{\rho}\left[\tau + \sigma\frac{\bar{E}}{h\nu} + \kappa\right] \tag{5.46}$$

as we have seen in Section 2.5. However, even though the two coefficients are approximately the same (except for high energies), Eq. (5.44) is the correct definition for B_a.

The exposure-dose buildup factor, B_r, is defined to be[6]

$$B_r(\bar{r}) = \frac{\int \left(\frac{\mu_a}{\rho}\right)_{air}(E)\, I(\bar{r}, E)\, dE}{\int \left(\frac{\mu_a}{\rho}\right)_{air}(E)\, I_0(\bar{r}, E)\, dE}$$

$$\equiv B_a^{air}(\bar{r}) \tag{5.47}$$

where $\mu_a/\rho)_{air}(E)$ is the absorption coefficient for air (that is, the detector is assumed to be an ideal air-wall ionization chamber). Goldstein and Wilkins[6] call this simply the dose buildup factor.

In order to calculate the buildup factors above, it is necessary to know $\phi(\bar{r}, E)$ or $I(\bar{r}, E)$. The question may then be raised about the need for buildup factors and uncollided flux densities if one must solve the exact problem anyhow. It turns out that the Boltzmann transport equation can be solved only for relatively simple geometries; however, one can improve on the uncollided flux density or dose rate estimates for more complex configurations by using these approximate buildup factors.

Several techniques have been developed in order to find solutions to the Boltzmann transport equation (5.35, or modifications of it), and we shall briefly discuss some of these techniques in the following paragraphs.

    a.   The "Straight-ahead" Approximation Method

    Basic assumptions are:

    1.   Infinite homogeneous absorbing medium

    2.   Neglect angular change in direction of photon; but, account for photon energy losses.

The second assumption is motivated by the fact that at high energies, Compton scatterings are predominately at small angles. This is seen, for example, in Fig. 4.5 of (FBM). The results of this approximation are quite poor (especially for low-Z materials) and may be off by several orders of magnitude.

b. Method of Successive Scatterings

In this technique, one makes use of the fact that the unscattered flux density is easily obtained. The unscattered flux density and the scattering cross section give the collision density (interactions/cm^3 sec) for first scattering. By treating such scattering collisions as new sources, the flux of singly-scattered photons can be found. A more detailed description of this technique is given by Goldstein.[4]

c. Monte Carlo Method

In this technique, each photon interaction is chosen at random from the given probability distribution for that type of process (such as absorption, scattering, etc.). One simply follows a sufficiently large number of photons through the medium, keeping record of their histories. The technique becomes prohibitive for deep penetrations due primarily to the length of time needed to perform the large number of calculations. The present generation of computers has alleviated this difficulty to some extent; however, modifications (approximations) of the basic Monte Carlo technique are generally necessary in order to make such calculations feasible. This is especially true for electromagnetic cascade shower calculations. The greatest asset of the Monte Carlo calculation lies in the fact that it can be applied essentially to any geometry. The Monte Carlo technique, however, does not solve the Boltzmann transport equation itself.

d. Method of Moments

This technique, which was originated by Spencer and Fano,[5] is a semi-numerical method for solving the Boltzmann transport equation. Except for distances significantly less than one mean free path from the source, the technique yields results of high accuracy for point and plane isotropic sources in infinite absorbing media. The basic shortcoming of this technique is governed by the above restrictions. In such cases, the method of successive scatterings and the Monte Carlo technique are superior. The authority on this subject is Goldstein and Wilkins[6] who plot the exposure-dose buildup factor (dose buildup factor in their nomenclature) and the absorbed-dose buildup factor (energy-absorption buildup factor in their nomenclature) as a function of the relaxation length, μr, for eight different media and for an energy range from 0.5 to 10 MeV. Some of their curves have been reproduced in the Appendix (Figs. A.25 through A.30).

The buildup factors of Goldstein and Wilkins[6] are the ones most notably used in the fields of dosimetry and health physics. A few precautionary remarks, are in order:

1. The exposure-dose buildup factor, B_r, (commonly called the dose buildup factor) is quite often mistaken as the quantity to be used for absorbed dose calculations, whereas one should really use the absorbed dose buildup factor, B_a, defined by Eq. (5.44) (the energy absorption buildup factor in Goldstein and Wilkins nomenclature).

2. Goldstein and Wilkins used μ_a/ρ in their calculations (Eq. (5.46)), whereas, to be precise they should have used μ_{en}/ρ (Eq. (5.45)). The difference is probably negligible but should be checked for each individual situation.

3. It should be reiterated that these calculations are for point and plane isotropic sources in infinite absorbing media.

5.13 Approximating the Buildup Factor with Formulas

As we have seen in previous sections, various equations can represent buildup factors. The most attractive representation, referred to as the Taylor formula, is given by

$$B(E_0, \mu r) = \sum_{n=1}^{N} A_n e^{-\alpha_n \mu r} \tag{5.48}$$

where E_0 is the primary-photon energy and μ is the associated linear attenuation coefficient. In practice, only two terms are required to fit the Goldstein-Wilkins[6] data to within 5% for the point isotropic source in an infinite medium. We have, therefore,

$$B(E_0, \mu r) = A_1 e^{-\alpha_1 \mu r} + A_2 e^{-\alpha_2 \mu r} \tag{5.49}$$

where

$$A_2 = 1 - A_1$$

and where A_1, α_1 and α_2 are functions of E_0 for any given medium. One has to decide on B_E, B_r, or B_a when choosing the parameters A_1, α_1 and α_2. These quantities are plotted in the Appendix (see Figs. A.31 through A.36).

The wisdom of fitting the Goldstein-Wilkins data to a sum of exponentials lies in the fact that one can, in many cases, replace the uncollided flux density equation by a sum of N terms (usually two), each identical in form to the uncollided-flux density equation but with fictitious attenuation coefficients given by $\mu(1+\alpha_n)$ and with weighting factors given by A_n. For example, the energy flux density, I, for a finite-line source in an infinite medium (with $\theta_1 = \theta_2 = \theta$) is given by the uncollided-flux density equation

$$I_0 = \frac{S_L E}{2\pi a} F(\theta, b)$$

where $b = \mu a$

a = source-to-detector distance

and by the energy flux density with buildup equation

$$I = \frac{S_L E}{2\pi a} \sum_{n=1}^{N} A_n F(\theta, b_n)$$

where

$$b_n = \mu(1+\alpha_n) a$$

Thus, the preceding calculations (Sections 5.4 through 5.10) for the uncollided-flux density can be taken over merely by invoking the above changes in the attenuation coefficient and weighting factor. Other formulas for approximating the buildup factor are given in (MT) and have been illustrated in Sections 5.5, 5.6 and 5.7.

In actual practice, one is faced with the problem of determining the buildup for a source-slab configuration, whereas the buildup factors discussed above are for sources in an infinite medium. An approximation most often used is simply to use the infinite medium buildup factors for such geometries. Comparisons with such calculations are readily found in the literature and are generally quite reasonable — at least for radiation protection applications.

When shielding consists of multiple layers, the problem of arriving at a highly accurate buildup factor becomes especially difficult. The crux of the problem of selecting a good buildup factor for such shielding arrangements lies in the fact that the flux incident on second and subsequent shielding layers is generally far different from that incident on the first layer. Therefore a product of buildup factors, one for each successive layer, is quite artificial unless the buildup factor for each layer is chosen on the basis of the energy flux incident on that layer. Since the energy flux on each layer beyond the first may be quite complex, the problem of generating a buildup factor is indeed formidable. At the present time no generally acceptable method of handling the problem is available. However, several empirical techniques for obtaining a buildup factor have been suggested. They should probably be thought of as rules of thumb generally yielding only rather rough predictions about flux and dose. Some of them are:

1. For a light material followed by a heavy material, only the buildup factor for the heavy material should be used.
2. For a heavy material followed by a light material, the product of the buildup factors is used (in the case of more than two slabs, this technique can be used but may yield a very conservative answer (i.e., flux and dose predictions on the high side)).
3. For a series of layers, the buildup factors entering into a product buildup factor may each be weighted according to the number of relaxation lengths of each shield material present.
4. The actual shield may be replaced by an equivalent shield of simple composition. "Equivalent" is used here in the sense of virtually identical in regard to gamma-penetration properties. (FBM)

5.14 Calculation of Absorbed Dose From Gamma Radiation

We have now reached a point where we can fully appreciate the complexity of and some of the difficulties associated with gamma ray dosimetry. The basic cause of the difficulties is the fact that not all gamma ray interactions are purely absorptive. This fact combined with various source geometries gives rise to the complex nature of gamma ray absorbed dose calculations. It is because of these complexities that an absolutely accurate calculation of absorbed dose in an object exposed to gamma radiation is virtually impossible.

Any object inserted into a radiation field will perturb that field by absorbing and scattering the gamma rays and electrons. We can, as a first approximation, assume that the object does not perturb the field and calculate the absorbed dose based on the calculated gamma ray flux density at the point when the object is not present. Assuming we have considered absorption and scattering in the flux density calculation, this approximation will generally be adequate when applied to small objects such as ion chambers or dosimeters used in dose measurement. The approximation is generally not adequate when one is interested in calculating the absorbed dose at some depth in a massive object such as a man. In this situation, attenuation and buildup should be considered since man is more than one mean free path thick (for $h\nu < 10$ MeV).

In this section we will discuss the equations necessary to calculate the gamma ray absorbed dose at a point assuming we have determined the flux density at that point by some method such as those detailed in 5.4 through 5.10. Consider, as a review, the following concepts and definitions that have been presented earlier:

Absorbed Dose — Absorbed dose (D) is the energy imparted per unit mass of an absorber.

Energy Imparted — Energy imparted is the sum of all energy entering a mass element on charged and uncharged particles minus the energy leaving the mass element on charged and uncharged particles minus the energy converted to rest mass in the mass element.

Mass Attenuation Coefficient — The mass attenuation coefficient measures the number of photons interacting (through any process) in passing through an absorbing medium.

$$\mu/\rho = \frac{1}{\rho}(\tau + \sigma + \sigma_R + \kappa)$$

The mass attenuation coefficient is generally used in the exponential when calculating the reduction in flux density of photons passing through an absorbing material (see Section 2.5).

Mass Energy Absorption Coefficient — The mass energy absorption coefficient measures the amount of energy deposited in a medium by photons interacting in the medium (see Section 2.5).

$$\mu_{en}/\rho = \frac{1}{\rho}\left[\tau(1-f) + \sigma\frac{\bar{E}}{h\nu} + \kappa\left(1 - \frac{2mc^2}{h\nu}\right)\right][1-G]$$

In principle, the calculation of absorbed dose is rather simple. One determines the photon flux density at the point of interest, multiplies by the energy of the photons to get the energy flux density and then by the mass energy absorption coefficient to determine how much of the energy is actually deposited at the point of interest. Finally, applying the appropriate constants to convert the units to rads and multiplying by the time during which the photon flux density was present yields the absorbed dose. Mathematically

$$D(\text{rads}) = 1.6 \times 10^{-8} \, \phi(\text{cm}^{-2} \text{sec}^{-1}) \, E(\text{MeV}) \frac{\mu_{en}}{\rho} (\text{cm}^2 \text{g}^{-1}) \, t(\text{sec}) \qquad (5.50)$$

In actual practice, however, the calculation of absorbed dose is generally very difficult and the best we can hope to do is obtain a reasonable approximation. We saw, in the sections above, how complex the calculation of the flux density becomes in all but the most simple geometrical situations. The addition of attenuators which introduce the need for scattering corrections compound the complexity. Scattering corrections using buildup factors are at best gross approximations, particularly since the buildup factors by nature of their determination are strictly applicable only in infinite media.

We must, in addition, account for the energy spectrum of the photons since in general the source will not be monoenergetic, and even if it is there will be an energy distribution after the photons have traversed an attenuating medium. In general, the energy dependence of the flux density, attenuation and energy absorption coefficients, and buildup factors are not easily written in an analytical form. Consequently, we are left with choosing an average or effective energy for the photons in our calculation and thus introducing another approximation.

Also, in calculating the absorbed dose by means of Eq. (5.50), we are assuming charged particle equilibrium at the point of interest, since the mass energy absorption coefficient treats only the energy deposited by photon interactions in the mass element at the point of interest. If charged particle equilibrium does not exist, we must somehow calculate the difference between energy entering and leaving the mass element on charged particles.

Finally, we must account for the fact that the flux density, and consequently the dose rate, may not be constant in time. If the source is a single radionuclide, the time variation of the flux density is determined by the half-life of the nuclide and is easily handled. However, radiation sources are seldom so simple and if the source is a combination of several radionuclides, fission products, or an operating reactor or accelerator, the treatment of the time variation of flux density (or dose rate) is rather complex.

An approximate formula that is often used to calculate the "dose" rate at 1 foot from a point isotropic gamma ray source is

$$R = 6 \, CE$$

where C is the source activity in curies and E is the gamma ray energy in MeV. The quantity that is actually calculated by means of this equation is the exposure rate in roentgens/hour. There are certain limitations to the use of this formula which should be understood, and the following derivation is useful in pointing out these limitations.

The flux density at 1 foot from a point isotropic source assuming no attenuation is

$$\phi = \frac{3.7 \times 10^{10} \, (\gamma\text{-sec}^{-1}\text{-Ci}^{-1}) \, 3.6 \times 10^{3} \, (\text{sec-hr}^{-1})}{4\pi \, (30.5 \text{ cm})^{2}} \, C(\text{Ci})$$

$$= 1.16 \times 10^{10} \, C \, (\gamma\text{-cm}^{-2}\text{-hr}^{-1})$$

In the energy region $0.07 < E < 2$ MeV the mass energy absorption coefficient for air is

$$\mu_{en}/\rho = 2.7 \times 10^{-2} \, \text{cm}^2\text{-g}^{-1} \, (\pm \, 15\%)$$

We will see that $1R = 87$ erg/g in air. Hence,

$$R = \frac{1.16 \times 10^{10} \, C \, (\gamma\text{-cm}^{-2}\text{-hr}^{-1}) \, 2.7 \times 10^{-2} \, (\text{cm}^{2}\text{-g}^{-1}) \, 1.6 \times 10^{-6} \, (\text{erg-MeV}^{-1})}{87 \, (\text{erg-g}^{-1}\text{-R}^{-1})} \, E(\text{MeV})$$

or

$$R(\text{roentgens/hr}) \simeq 6 \, CE \qquad (5.51)$$

where C is the activity in Curies and E is the photon energy in MeV. Thus, in the energy range $0.07 < E < 2.0$ MeV this formula can be expected to give the exposure rate (to within $\sim 20\%$) at 1 foot from a point isotropic gamma source, assuming no attenuation or buildup.

The relationship between exposure and absorbed dose is another important concept. The importance of the relationship will become more evident in Chapter 6 when we discuss dose measurements. What is generally measured is exposure and an understanding of the relationship of exposure to absorbed dose is necessary.

If we make use of the terms already defined:

Particle Fluence	ϕ
Energy Fluence	F
Absorbed Dose	D
Exposure	X
Mass Energy Absorption Coefficient	μ_{en}/ρ
Mass Stopping Power	$\frac{1}{\rho}\frac{dT}{dx}$

we can develop certain relationships between them in the calculation of absorbed dose. First, we introduce the quantity, W. W is the energy required to produce one ion pair in air and has a measured value of 34 eV/i.p. for most radiations and energies of interest. Using this quantity we can calculate the absorbed dose in air exposed to 1R under charged particle equilibrium.

$$D = \frac{2.58 \times 10^{-4} \, (C\text{-}kg^{-1}) \; 34 \, (eV\text{-}ip^{-1}) \; 1.6 \times 10^{-12} \, (erg\text{-}eV^{-1})}{1.6 \times 10^{-19} \, (C\text{-}ip^{-1}) \; 1 \times 10^{3} \, (g\text{-}kg^{-1}) \; 1 \times 10^{2} \, (erg\text{-}g^{-1}\text{-}rad^{-1})}$$

$$= 0.87 \text{ rad}$$

In general then, the absorbed dose in air is given by

$$D\,(rad) = 0.87 \, X\,(roentgen) \tag{5.52}$$

Now, if we have a monoenergetic photon beam of energy E, the energy fluence is $F = \phi E$. With E measured in ergs the absorbed dose at a point in air will be given by

$$D\,(rad) = (1/100)\phi E \, (\mu_{en}/\rho)_{air} = 0.87 \, X\,(roentgen) \tag{5.53}$$

from above.

If the beam of photons has a spectrum with a maximum energy E_m, then the absorbed dose is given by

$$D(\text{rad}) = \frac{1}{100} \int_0^{E_m} \phi(E) \left(\frac{\mu_{en}}{\rho}\right)_{air} E \, dE \qquad (5.54)$$

where $\phi(E)$ now has the units cm^{-2} MeV^{-1}.

If the medium involved is not air and charged particle equilibrium exists, then the dose to the medium is

$$D_M(\text{rad}) = 0.87 \, X \, \frac{(\mu_{en}/\rho)_M}{(\mu_{en}/\rho)_{air}} \qquad (5.55)$$

where X is exposure in roentgen.

Up to now we have considered photons as the particle incident on the medium of interest. If the particles are charged particles with a fluence per unit energy interval $\phi(E)$ entering a volume of cross section area dA and depth $d\ell$, the dose is

$$D(\text{rad}) = \frac{1.6 \times 10^{-8} \int_0^{E_m} \frac{dT}{dx}(E) \, \phi(E) \, dA \, d\ell \, dE}{\rho \, dA \, d\ell}$$

$$= \frac{1.6 \times 10^{-8}}{\rho} \int_0^{E_m} \frac{dT}{dx}(E) \, \phi(E) \, dE \qquad (5.56)$$

where the stopping power, dT/dx, has the units $MeV - cm^{-1}$.

REFERENCES

1. Reactor Shield Design Manual, Theodore Rockwell III, editor (D. Van Nostrand Co., Inc., Princeton, N. J., 1956).
2. Engineering Compendium on Radiation Shielding, Vol. I, "Shielding fundamentals and methods" (Springer-Verlag, New York, 1968).
3. J. J. Taylor and F. E. Obershain, USAEC Report WAPD-RM-213, Westinghouse Electric Corp. (1953).
4. H. Goldstein, Fundamental Aspects of Reactor Shielding (Addison-Wesley Publishing Co., Inc., Reading, Mass., 1959).
5. L. V. Spencer and U. Fano, Phys. Rev. $\underline{81}$, 464 (1951); J. Res. Natl. Bur. Std. $\underline{46}$, 446 (1951).
6. H. Goldstein and J. E. Wilkins, Jr., Calculations of the Penetration of Gamma Rays, USAEC Report NYO-3075, Nuclear Development Associates, Inc. (1954).

MAIN REFERENCES

(FBM) J. J. Fitzgerald, G. L. Brownell, and F. J. Mahoney, Mathematical Theory of Radiation Dosimetry (Gordon and Breach, New York, 1967).

(MT) K. Z. Morgan and J. E. Turner (ed.), Principles of Radiation Protection (Krieger Publishing Co., New York, 1973).

CHAPTER 6

MEASUREMENT OF RADIATION DOSE — CAVITY-CHAMBER THEORY

6.1 Introduction

To measure absorbed dose (energy absorbed per unit mass) in a medium exposed to ionizing radiation one must introduce into the medium a radiation sensitive device. Normally, this device will constitute a discontinuity in the medium since it generally differs from the medium in atomic number and density. Because of these differences we know from the previous chapters that it will have different properties with regard to absorption of energy from ionizing radiations. This radiation sensitive device can be a gas, liquid, or solid and will be referred to as a cavity.

Consider this cavity situated in a medium permeated by a spatially uniform flux density of photons (ϕ). At any point within this medium (at a depth equal to or greater than the maximum secondary electron range*), charged particle equilibrium will be closely approximated and the photon flux density will give rise to a spatially uniform electron flux density (ϕ_e). By considering a finite exposure time t and defining fluence $\Phi = \phi t$ (or $\Phi_e = \phi_e t$) we can determine the absorbed dose to the medium** (M):

$$D_M = \Phi E \, (\mu_{en}/\rho)_M \, .$$

This can also be written, using the electron fluence

$$D_M = \Phi_e \left(\frac{1}{\rho} \frac{dT}{dx} \right)_M$$

---

*Note: secondary electrons are those electrons produced by photons; knock-on electrons from these secondary electrons will be called δ-rays.

**We assume throughout this discussion that G=0, so that $\mu_{en}/\rho = \mu_K/\rho$.

where it is understood that

$$\Phi E \frac{\mu_{en}}{\rho} = \int_0^{E_{max}} \frac{d\Phi(E)}{dE} E \frac{\mu_{en}}{\rho}(E) \, dE$$

and

$$\Phi_e \frac{1}{\rho}\frac{dT}{dx} = \int_0^{E_{max}} \frac{d\Phi_e(E)}{dE} \frac{1}{\rho}\frac{dT}{dx}(E) \, dE \; .$$

Now, if we introduce a cavity into this medium, the absorbed dose to the cavity will in general be different from the absorbed dose to the medium. The relationship between the dose to the cavity and the dose to the medium depends on the cavity material and the cavity size. In general, we will assume the cavity material is different from the medium. Concerning cavity size, there are three situations.

1. Cavity dimensions small compared with the electron range.
2. Cavity dimensions large compared with the electron range.
3. Cavity dimensions of the order of the electron range.

The first situation was assumed in the development of the Bragg-Gray theory. However, later theories by Laurence, Spencer and Attix, Burch, and Burlin have allowed the extension of the Bragg-Gray theory to situations 2 and 3.

6.2 Cavity Size Small Relative to Range of Electrons

A. Basic Assumptions

The requirements underlying the statement that the cavity size is small relative to the range of the electrons imply the following assumptions (ART):

1. The secondary electron spectrum generated in the medium by the primary photon flux density is not modified by the presence of the cavity material.

2. Photon interactions which generate secondary electrons in the cavity can be neglected.
3. The primary photon fluence in the region from which secondary electrons can enter the cavity is spatially uniform. This implies that the secondary electron fluence (ϕ_e) is also uniform.

B. **Bragg-Gray Cavity Theory**

We assume, as Gray did,[1] that the introduction of a gas-filled cavity into a homogeneous medium does not change the electron spectrum that is present in the medium. In other words,

$$\phi_e^C = \phi_e^M = \phi_e$$

where ϕ_e is the electron fluence (which could have been written as a differential, $d\phi_e/dE$ as well).

Consider now, two geometrically identical volume elements — to make it easier (but less general), two cubes — one a small cavity in an irradiated medium and the other a solid element of the uniformly irradiated medium. Let the respective linear dimensions of the two volume elements be in the ratio s:1, where*

$$s = \frac{\frac{dT}{dx}\big|_M \text{ (MeV/cm)}}{\frac{dT}{dx}\big|_C \text{ (MeV/cm)}} \qquad (6.1)$$

so that the volume elements are related by

$$\delta V_C = s^3 \, \delta V_M$$

Let δE be the amount of energy lost by one electron in crossing the volume, δA be the cross-sectional area of an element and δN be the number of electrons

---

*s is called the stopping power ratio.

crossing the volume. Then,

$$\delta E_C = \frac{dT}{dx}_C \cdot s$$

$$\delta E_M = \left(\frac{dT}{dx}\right)_M \cdot 1$$

This leads to $\delta E_C = \delta E_M$. Also

$$\delta N_C = \phi_e^C \, \delta A_C = \phi_e \, s^2$$

$$\delta N_M = \phi_e^M \, \delta A_M = \phi_e$$

which leads to $\delta N_C = s^2 \, \delta N_M$. Hence, if $_v E$ denotes the energy lost per unit volume, we have

$$_v E_C = \frac{\delta N_C \, \delta E_C}{\delta V_C} = \frac{s^2 \, \delta N_M \, \delta E_M}{s^3 \, \delta V_M}$$

But

$$_v E_M = \frac{\delta N_M \, \delta E_M}{\delta V_M}$$

so that

$$_v E_C = \frac{1}{s} \, _v E_M$$

That is, the energy lost (per cm^3) by electrons in the cavity is $1/s$ times that lost in the medium. The basic assumption here is that ϕ_e (or $d\phi_e/dE$) is unchanged — in other words, the cavity is small relative to the range of the electrons and the electron energy loss is continuous.

Now, we have seen (Chapter 1) that the energy imparted to matter by electrons in the mass element dm is

$$E_D = (\textstyle\sum E_E)_c - (\textstyle\sum E_L)_c + (\textstyle\sum E_E)_u - (\textstyle\sum E_L)_u - (\textstyle\sum E_R)_u$$

(here c = charged particle, u = uncharged particle) and that under charged particle equilibrium conditions

$$(\sum E_E)_c = (\sum E_L)_c$$

by definition. Thus

$$E_D = (\sum E_E)_u - (\sum E_L)_u - (\sum E_R)_u$$
$$= E_K$$

so that the energy imparted (i.e., lost) by the secondary electrons in a volume (mass) element in the medium is equal to the energy lost by the photons through interactions within that volume (mass) element (assuming G = 0; that is, bremsstrahlung production is negligible).

We can now state Gray's principle of equivalence from the above two statements:

"The energy lost per unit volume by electrons in the cavity is 1/s times the energy lost by γ-rays per unit volume of the solid." (ART)

To complete the derivation of the Bragg-Gray relation, we must now make a further assumption, as Gray[2] did, that energy lost by the electrons in crossing the volume is equal to the energy deposited in the volume for both cavity and medium. In other words, energy does not leave the volume in the form of δ-rays without being replaced by an equivalent amount of energy entering.

Now, if $_vJ$ is the ionization per unit volume of gas, and if the average energy dissipated in the gas per ion pair formed, W, is independent of energy, we can calculate the energy absorbed per unit volume of the solid by

$$_vE_M = s\ _vE_C = sW\ _vJ \qquad (6.2)$$

which is called the Bragg-Gray formula.

It is more common to use the energy absorption per unit mass in the solid, $_mE_M$, and the ionization per unit mass in the gas, $_mJ$, which comes about from the above equation as follows, where the m denotes mass:

$$_mE_M \, \rho_M = sW \, _mJ \, \rho_C$$

But, we let

$$_ms = \frac{\left(\frac{1}{\rho}\frac{dT}{dx}\right)_M}{\left(\frac{1}{\rho}\frac{dT}{dx}\right)_C} \tag{6.3}$$

to get

$$_mE_M = \, _ms \, W \, _mJ \tag{6.4}$$

C. Extensions of the Bragg-Gray Theory

In addition to the assumptions stated above, Gray also concluded that the stopping power ratio was "almost independent of the energy of the electrons". In reality, it is not and Laurence (1937) modified the Bragg-Gray theory to account for the energy dependence of the stopping power ratio (ART). By assuming a continuous energy loss model for electrons traversing a medium the secondary electron spectrum is given by $\left(\frac{1}{\rho}\frac{dT}{dx}\right)_M^{-1}$ at CPE (i.e., the reciprocal of the mass stopping power of the medium). Under these conditions Laurence derived an expression for the mass stopping power ratio of the medium to the cavity gas (subscripts Z and G, respectively):

$$\frac{1}{_ms} = \frac{(Z/A)_G}{(Z/A)_Z}\left[1 + b_Z(T_0)\frac{I_Z}{I_G} + d_Z(T_0)\right] \tag{6.5}$$

In this equation, $b_Z(T_0)$ and $d_Z(T_0)$ are functions of the initial electron energy and have been tabulated in NBS Handbook 79.[3] In addition $b_Z(T_0)$ depends to a small extent on the ionization potentials (I_Z and I_G). The function $d_Z(T_0)$ accounts for the density (polarization) effect.

The inherent assumption in the Bragg-Gray theory that the electron energy loss is continuous is also not strictly correct. In 1955, Spencer and Attix and in an independent paper, Burch, published theories to account for the discrete energy losses by electrons (ART). The Spencer-Attix theory limited the stopping power ratio to energy losses below an arbitrary energy limit Δ. In practice, Δ is taken to be the energy of an electron which will just cross the cavity. Consequently, Δ is not only energy dependent but also dependent on the cavity size (or gas pressure). Burch used the same model as Gray but redefined his volume element dimensions to exclude the energy leaving the volume on δ-rays or bremsstrahlung. The extreme difficulties involved in this formulation have prevented any numerical solution to the theory.

Spencer and Attix were able to derive an approximate expression for the ratio of the total electron flux density to the primary electron flux density at an energy T for electrons of initial energy T_0. This expression $\{R_Z(T_0, T)\}$ is easily obtained numerically and is used to calculate the fast electron spectra

$$K_Z(T_0, T) = R_Z(T_0, T) \left(\frac{1}{\rho} \frac{dT}{dx}\right)_Z^{-1}$$

The result is that Spencer and Attix were able to derive an analytical expression for the mass stopping power ratio taking into account both the energy dependence and the fact that the electrons do not lose energy continuously. The formula is given in the same form as the Laurence formula in NBS Handbook 79[3]:

$$\frac{1}{{}_m s} = \frac{(Z/A)_G}{(Z/A)_Z} \left\{1 + c_Z(T_0, \Delta) \ln \frac{I_Z}{I_G} + d_Z(T_0)\right\} \tag{6.6}$$

Again, the functions $c_Z(T_0, \Delta)$ and $d_Z(T_0)$ are tabulated. The cavity size dependence enters through $c_Z(T_0, \Delta)$ while $d_Z(T_0)$ is identical to the $d_Z(T_0)$ in the Laurence equation.

These modifications to the basic Bragg-Gray stopping power ratio are important for certain situations, in particular, when charged particle equilibrium does not exist. This situation may arise at the interface between the medium and the cavity or when the primary photons have energies greater than a few MeV. When this occurs, there will be an imbalance between the energy entering the volume and the energy leaving the volume on charged particles. Hence, the Bragg-Gray assumption that the energy lost in the volume by secondary electrons is equal to the energy lost by photons through interactions in the volume is no longer valid. That is:

$$E_D^M = (\sum E_E^M)_c - (\sum E_L^M)_c + (\sum E_E^M)_u - (\sum E_L^M)_u - (\sum E_R^M)_u$$

and

$$(\sum E_E^M)_c \neq (\sum E_L^M)_c$$

However, the result $_vE_C = \frac{1}{s}\,_vE_M$ or identically $E_D^C = \frac{1}{s} E_D^M$ is still valid. That is, the energy imparted to the cavity is related to the energy imparted to the medium by $1/s$. Since the absorbed dose is defined in terms of energy imparted it will still be measured properly by the cavity provided the correct value is chosen for s.

In general, in the energy region where CPE can not be assumed, we can also not assume δ-ray equilibrium. Consequently, the energy lost in the cavity is not necessarily equal to the energy deposited in the cavity. This is the situation the Spencer-Attix theory attempts to take into account by choosing a limit on the amount of energy lost which can still be considered locally deposited. In fact, what is done is to use a restricted stopping power ratio in place of $_m s$. The energy restriction is based on the cavity size. Thus we can write:

$$_m E_M = {}_m^{\Delta} s \, W \, {}_m J$$

At higher energies, the secondary electrons may also lose energy by bremsstrahlung production. In this case, the energy lost is most certainly not deposited locally. Consequently, one must realize that the correct stopping power to use is the collision stopping power. This will differ greatly from total stopping power (i.e., collision plus radiation loss) at high energies.

The effect of the Spencer-Attix modifications is shown in Fig. 6.1. It is obvious that the consideration of cavity size is important only for grossly mismatched media such as lead and air. In a well-matched system, δ-ray equilibrium may exist and the Laurence formulation for stopping power may be adequate. Whereas the Spencer-Attix formulation must be used, when the system is significantly mismatched.

6.3 The Effect of Cavity Size[4,5]

We have discussed in detail the theoretical development for absorbed dose measurements using a small cavity. The qualitative effect of cavity size is shown in Fig. 6.2.

A. Small Cavity (Fig. 6.2B)

In this situation, the cavity is small enough that the electron fluence is not perturbed by the cavity. Also, there is no appreciable photon interaction in the cavity. Thus, the absorbed dose expressions are:

$$D_M = \Phi E \, (\mu_{en}/\rho)_M = \Phi_e^M \left(\frac{1}{\rho} \frac{dT}{dx}\right)_M \quad *$$

---

*Note: As in Section 6.1 $\Phi E \dfrac{\mu_{en}}{\rho} = \displaystyle\int_0^{E_{max}} \dfrac{d\Phi(E)}{dE} E \dfrac{\mu_{en}}{\rho}(E) \, dE$ and $\Phi_e \dfrac{1}{\rho}\dfrac{dT}{dx} = \displaystyle\int_0^{E_{max}} \dfrac{d\Phi_e(E)}{dE} \dfrac{1}{\rho}\dfrac{dT}{dx}(E) \, dE$. Thus $(\mu_{en}/\rho)_M/(\mu_{en}/\rho)_C$ and $_m s$ are average values taken over the appropriate energy spectrum.

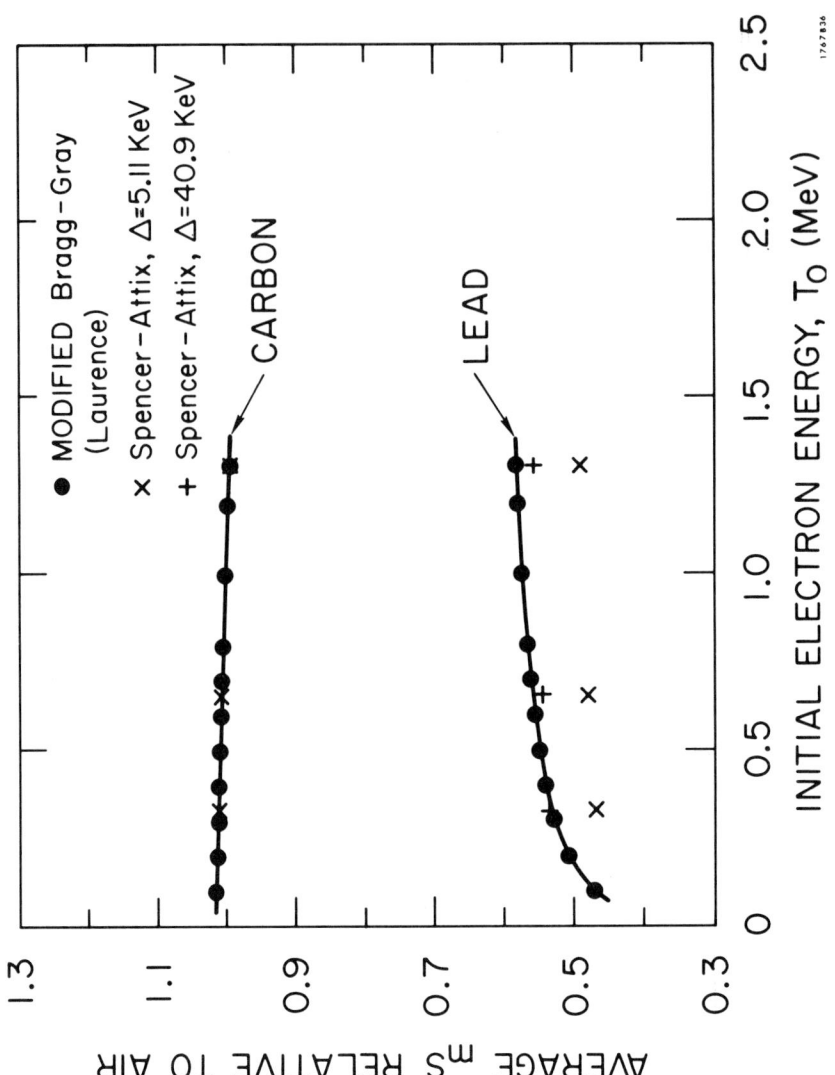

FIG. 6.1 Stopping power ratios in carbon and lead using various theoretical treatments (data from Attix et al., Vol. I).

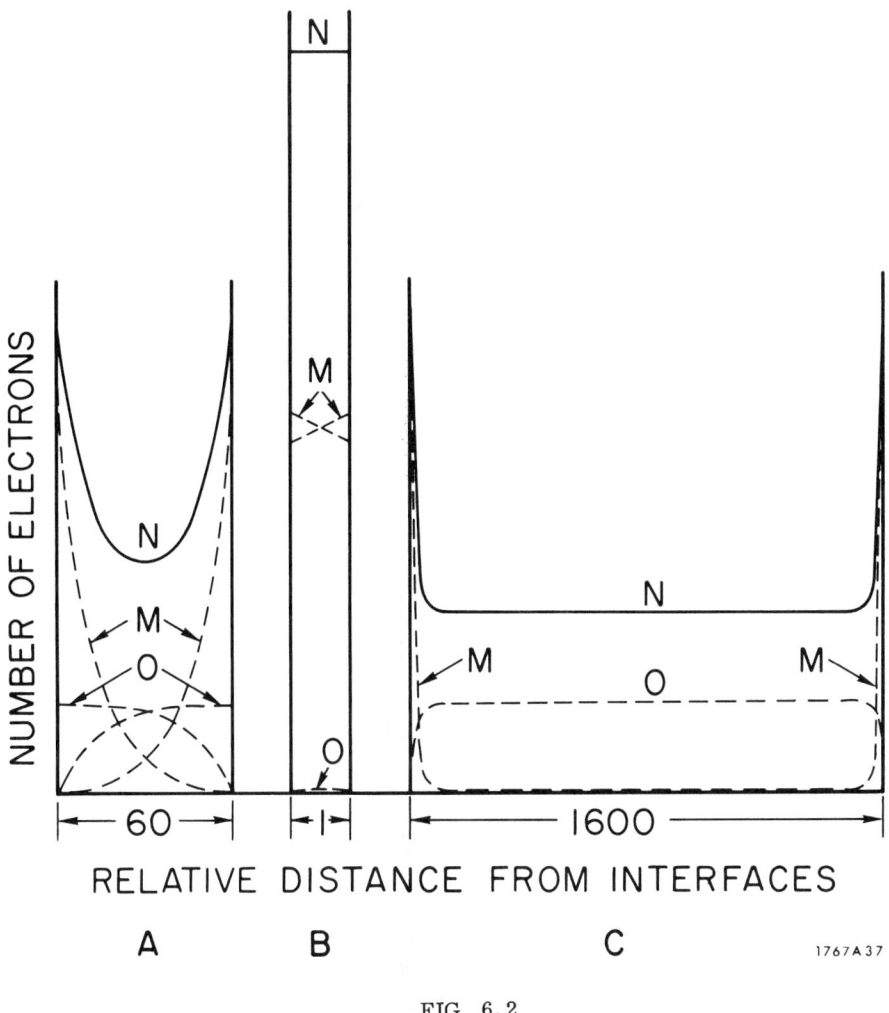

FIG. 6.2

Electron distributions in various size cavities
 M - electrons entering the cavity from the medium
 O - electrons generated by photon interactions in the cavity
 N - total number of electrons.

for the medium and

$$D_C = \Phi_e^C \left(\frac{1}{\rho}\frac{dT}{dx}\right)_C$$

for the cavity. In this case, the electron fluence is the same in the medium and the cavity and so, $\left(\Phi_e^M = \Phi_e^C\right)$

$$D_M/D_C = \left(\frac{1}{\rho}\frac{dT}{dx}\right)_M / \left(\frac{1}{\rho}\frac{dT}{dx}\right)_C$$

$$D_M/D_C = {}_m s$$

where ${}_m s$ is defined as the relative mass stopping power of the medium to the cavity. Figure 6.2 illustrates the case where ${}_m s$ is greater than one but this need not be the case in general.

B. <u>Cavity Size Large Relative to Range of Electrons (Fig. 6.2C)</u>

When the cavity dimensions are many times larger than the range of the most energetic electrons produced in the medium, the contribution to the absorbed energy in the cavity from the region of the medium/cavity interface is negligible. Thus the energy absorbed in the cavity will depend only on the cavity material. Similarly, the energy absorbed in the medium will depend only on the properties of the medium, except in the immediate region of the interface.

If we consider the dose at points greater then the electron range from the interface, we arrive at the following dose expressions: $D_M = \Phi E(\mu_{en}/\rho)_M$ for the medium and $D_C = \Phi E(\mu_{en}/\rho)_C$ for the cavity. Assuming the dimensions of the medium and cavity are still small enough so that ΦE does not change appreciably

$$D_M/D_C = (\mu_{en}/\rho)_M / (\mu_{en}/\rho)_C$$

Figure 6.2 illustrates the case where $(\mu_{en}/\rho)_M$ is greater than $(\mu_{en}/\rho)_C$ but this need not be the case in general. At the interface between the medium and the cavity there will be a discontinuity in the absorbed dose because of the difference in the scattering properties and stopping powers of the two materials. We can also write the absorbed dose using the electron fluence as

$$D_M = \Phi_e^M \left(\frac{1}{\rho}\frac{dT}{dx}\right)_M$$

$$D_C = \Phi_e^C \left(\frac{1}{\rho}\frac{dT}{dx}\right)_C$$

In general, $\Phi_e^M \neq \Phi_e^C$ in this situation even though the photon fluence is unperturbed.

C. Cavity Size Comparable to Range of Electrons (Fig. 6.2A)

When the cavity size is comparable to the electron range, the first two assumptions of small cavity theory (Section 6.2A) are no longer valid. The secondary electron spectrum generated in the medium (or cavity wall) is modified within the cavity, and secondary electrons generated within the cavity by photon interactions become important. On the other hand, the region of interface between the cavity and the medium is no longer negligible as it was in the large cavity case. This situation has been treated by Burlin through a slight modification to the Spencer-Attix equation for calculating mass stopping power ratios.

This modification to the theory for small cavities is based on the results of measurements made using a parallel plate extrapolation chamber to determine the effect of cavity size on ionization per unit mass of air in the cavity. The modification allows the mass stopping power ratio formula to approach the Spencer-Attix formula for small cavities while for large cavities it approaches the mass energy absorption coefficient ratio. The correction is most important when the difference between the atomic numbers of the medium and cavity gas is large and the value of Δ (electron energy cutoff) is large. For small Δ and well-matched cavities, the correction is negligible.

The analytical expression for the mass stopping power ratio developed by Burlin is (ART):

$$\frac{1}{{}_m s} = \frac{(Z/A)_G}{(Z/A)_Z} \left\{ 1 + d \left[c_Z(T_0, \Delta) \ln \frac{I_Z}{I_G} + d_Z(T_0) \right] \right.$$

$$\left. + (1-d) \left[\frac{(\mu_{en}/\rho)_G}{(\mu_{en}/\rho)_Z} \frac{(Z/A)_Z}{(Z/A)_G} - 1 \right] \right\} \quad (6.7)$$

The factor d is based on the well-verified exponential attenuation of electrons and is given by:

$$d = \int_0^t e^{-\beta x} \, dx \Big/ \int_0^t dx = \frac{1}{\beta t} (1 - e^{-\beta t})$$

where β is the effective electron attenuation coefficient and d=1 corresponds to a cavity size (t) approaching zero while d=0 corresponds to a cavity size (t) approaching infinity.

Using the mass stopping power ratio calculated in this manner allows the use of cavity chamber theory irrespective of cavity size, CPE or δ-ray equilibrium.

6.4 Measurement of Absorbed Dose (ART, MT and Ref. 6)

Absorbed dose measurements using cavity chamber theory can be made under a number of different conditions. These include gas ionization chambers with and without matched gas and wall material, ionization chambers calibrated for exposure, and devices other than ionization chambers. In this section we will briefly discuss absorbed dose measurements under these various conditions.

A. **Matched Gas and Wall Material**

This is a special case and its particular usefulness arises because of a theorem rigorously proved by Fano (1954) and stated as follows by Failla

(1956) (ART):

"In a medium of given composition exposed to a uniform flux of primary radiation, the flux of secondary radiation is (1) uniform, (2) independent of the density of the medium, and (3) independent of density variations from point-to-point, provided that the interactions of the primary radiation and the secondary radiation with the atoms of the medium are both independent of density."

This means that for a cavity in which the walls are of the same material as the cavity gas the mass stopping power ratio is unity regardless of the cavity size or the gas pressure, provided the density (polarization) effect is negligible. In principle, then, the Bragg-Gray condition that the cavity must be small compared with the electron ranges can be relaxed.

In practice, however, it is not easy to exactly match a cavity wall and gas in atomic composition. It can be done using ethylene in polyethylene or acetylene in polystyrene for example. Several approximations to air equivalent walls have been made generally using a bakelite/graphite mixture. An exact match requires identical mass energy absorption coefficients as well as identical mass stopping powers for the wall and gas. Recalling from the discussions in Chapters 2 and 3, the dependence on Z and A of μ_{en}/ρ is in general different from $\frac{1}{\rho}\frac{dT}{dx}$ and consequently matching one will result in a mismatch in the other. Finally, the density effect is seldom negligible at energies above a few MeV.

If we assume a cavity with perfectly matched walls and gas (e.g., an air cavity with air walls in an air medium), $_m s = 1$ and the absorbed dose would be (Eq. (6.4)):

$$D = 100\ _m E = 100\ W\ _m J$$

where W is the energy absorbed per unit charge (joules/Coul) and $_m J$ is the measured ionization per unit mass in the cavity gas (Coul/kg). If the photon field is equal to one roentgen, $_m J = 2.58 \times 10^{-4}$ Coul/kg and D = 0.87 Rad for an air cavity under CPE conditions.

Now if this same cavity is placed in a medium other than air, but the cavity wall is thick enough to ensure that only electrons originating in the wall enter the cavity, the absorbed dose measured will be the absorbed dose in the cavity wall. To arrive at the absorbed dose in the medium we must apply an additional condition. The ion chamber must be calibrated for the photon spectrum existing in the medium. If it is not, a perturbation correction must be made.[4] Assuming the chamber has been calibrated in roentgens, the absorbed dose ratio is:

$$\frac{D_M}{D_{air}} = \frac{(\Phi E)_M (\mu_{en}/\rho)_M}{(\Phi E)_{air} (\mu_{en}/\rho)_{air}}$$

or

$$\frac{D_M}{0.87 X} = \frac{(\mu_{en}/\rho)_M}{(\mu_{en}/\rho)_{air}}$$

since $(\Phi E)_M = (\Phi E)_{air}$. So, the dose to the medium will be

$$D_M = 0.87 \, X \left[\frac{(\mu_{en}/\rho)_M}{(\mu_{en}/\rho)_{air}}\right] \text{ (rad)} \tag{6.8}$$

when the chamber records an exposure of X roentgens.

The mass energy absorption coefficient ratio arises because the air cavity measures electrons generated by photon interactions in the air wall while the absorbed dose to the medium is delivered by electrons generated by photon interactions in the medium.

B. **Wall Material Different from Cavity Gas**

When the cavity wall material is not matched with the cavity gas, two situations can occur. Either the wall can be composed of the irradiated medium in which the measurement is being made, or the medium, wall and gas can be different materials.

In the first situation the absorbed dose to the medium is given by:

$$\frac{D_M}{D_C} = \frac{\Phi_e^M \left(\frac{1}{\rho}\frac{dT}{dx}\right)_M}{\Phi_e^C \left(\frac{1}{\rho}\frac{dT}{dx}\right)_C}$$

$$D_M = D_C \,_m s \tag{6.9}$$

where $_m s$ is the mass stopping power ratio of the medium to the cavity gas and the differences between Φ_e^M and Φ_e^C have been accounted for in the calculation of $_m s$.

In the second situation we must consider the difference in photon interactions between the medium and cavity wall in addition to the difference in stopping power between the medium and the cavity gas. Thus the absorbed dose to the medium is:

$$D_M = D_W \frac{(\mu_{en}/\rho)_M}{(\mu_{en}/\rho)_W} \qquad \text{from Eq. (6.8)}$$

The absorbed dose to the wall is:

$$D_W = D_C \,_m s^{wall}_{cavity} \qquad \text{from Eq. (6.9)}$$

Thus

$$D_M = D_C \,_m s^W_C \frac{(\mu_{en}/\rho)_M}{(\mu_{en}/\rho)_W} \tag{6.10}$$

If the cavity gas is air,

$$D_M = 0.87 \times {}_m s^W_{air} \frac{(\mu_{en}/\rho)_M}{(\mu_{en}/\rho)_W}.$$

where all the terms are defined as before and $(\mu_{en}/\rho)_W$ is the mass energy absorption coefficient for the wall material.

We must re-emphasize that the above equations (6.8, 9, 10) apply under all conditions only when ${}_m s$ properly includes the effect of discontinuous energy loss by electrons and the electron energy spectrum and cavity size have been accounted for (Burlin formulation).

C. Devices other than Ionization Chambers

Although much of the preceding discussion has referred to the cavity in terms of a gas-filled ionization chamber, cavity theory is general and can be applied to any cavity material. It is necessary only to insure that the cavity is small relative to the electron range, or apply the modified theory for larger cavities. For an air cavity at 1 atm pressure a small cavity for 1 MeV photons would be 1 cm or less. A solid or liquid cavity should have linear dimensions smaller than this by the ratio of the densities; that is, a unity density cavity should be 10^{-3} cm or less for the above situation.

When the cavity and its wall are of the same material, the absorbed dose to the medium is

$$D_M = D_C \left[\frac{(\mu_{en}/\rho)_M}{(\mu_{en}/\rho)_C} \right] \qquad \text{from Eq. (6.8)}$$

When the cavity wall and medium are of the same material, the absorbed dose to the medium will be given by

$$D_M = {}_m s \, D_C \qquad \text{from Eq. (6.9)}$$

where $_m s$ is the appropriate mass stopping power ratio of the medium to the cavity material. If the cavity material must be contained in some material different from the cavity material or the medium, we must take account of the differences in photon absorption between the medium and the cavity wall as before

$$D_M = D_C \, _m s \, \frac{(\mu_{en}/\rho)_M}{(\mu_{en}/\rho)_W} \qquad \text{from Eq. (6.10)}$$

The quantity D_C in the above expressions is the absorbed dose measured in the cavity material. This, of course, must be related to some response of the cavity material through an appropriate calibration.

When small well-matched cavities can be achieved, the simpler formulations for $_m s$ can be used. However, the cavity size limitation can be troublesome in practice for solid dosimeters and low energy photons. Recent work[7] indicates that for TLD materials the response for energies below 0.2 MeV is very dependent on the grain size of the TL material and thus the more complex formulation of $_m s$ is required. At higher energies, of course, a cavity size small with respect to the range of secondary electrons is easier to achieve.

6.5 Applications of Cavity Theory for Photon Fields (f and C_λ)

Most often for low and intermediate energy photons one will use an air cavity with air-equivalent walls. In this case Equation 6.8 expresses the relevant relationship and the value of $0.87 \, \frac{\mu_{en}/\rho)_M}{\mu_{en}/\rho)_{air}}$ is called the f-factor. The f-factor converts exposure measured in an air cavity with air-equivalent walls to dose in a medium.

To arrive at the dose using an air cavity chamber generally requires the measurement of exposure because national standards laboratories provide calibrations in terms of exposure in air. The exposure at any point in a medium will be given by

$$X = R \cdot N_c \cdot k_{TP} \cdot k \cdot d_c \tag{6.11}$$

where

- R is the chamber reading;
- N_c is the exposure calibration;
- k_{TP} is the temperature-pressure correction for an unsealed chamber;
- k corrects for differences in the radiation field between calibration and use (usually neglected);
- d_c is the displacement factor which corrects for the fact that the cavity displaces some of the medium.

The displacement factor corrects for absorption in and scatter from the medium displaced by the cavity. It is a function of chamber size and photon energy and is generally less than unity (about 0.985 for a 6mm diameter thimble chamber with buildup at Co-60 energies).

For typical medical radiological physics applications an air cavity with air-equivalent walls is used to measure dose in water and from equation 6.8 we have, at the energy for which N_c is determined

$$D_w = 0.87 \, X_w \, \frac{(\mu_{en}/\rho)_w}{(\mu_{en}/\rho)_a}$$

$$= 0.87 \, R \cdot N_c \cdot k_{TP} \cdot k \cdot d_c \, \frac{(\mu_{en}/\rho)_w}{(\mu_{en}/\rho)_a} \tag{6.12}$$

The quantity $0.87 \, d_c \, \frac{(\mu_{en}/\rho)_w}{(\mu_{en}/\rho)_a}$ is denoted $(C_\lambda)_c$, the conversion factor from chamber reading to dose at the energy for which the chamber is calibrated. The factor $(C_\lambda)_c$ is generally applied to cylindrical chambers calibrated at Co-60 energies or 2 MV bremsstrahlung. Note that the displacement factor is included in $(C_\lambda)_c$ and consequently the value of $(C_\lambda)_c$ will depend on the geometry of the cavity chamber. For parallel plate chambers with small dimensions, d_c is unity and $(C_\lambda)_c = f$ at the energy for which the chamber is calibrated.

If a chamber is used for measurements at an energy different from that at which it is calibrated, further corrections are required. For low and intermediate energies the value of N_c is usually determined by extrapolating between calibration energy points, but for energies above Co-60 or 2 MV this is not possible. In defining the exposure to dose conversion factor C_λ for measurements made in water at energies above Co-60, the assumption is made that electrons causing ionization in the cavity are generated in the water outside the cavity and not in the cavity wall. Consequently, the critical parameter in determining the response of the chamber at higher energies is the stopping power ratio of water to air (the cavity gas). Under this condition the factor C_λ is given by

$$C_\lambda = (C_\lambda)_c \frac{\left(s^w_{air}\right)_\lambda}{\left(s^w_{air}\right)_c} \frac{p_\lambda}{p_c} \qquad (6.13)$$

where

s^w_{air} indicates stopping power ratio of water to air;

λ indicates the energy of the photons;

$\dfrac{p_\lambda}{p_c}$ is the ratio of perturbation correction factors at the two energies and is generally assumed to be unity.

Then the dose to water in a photon field of energy λ will be

$$D_w)_\lambda = R \cdot N_c \cdot k_{TP} \cdot k \cdot C_\lambda \ . \qquad (6.14)$$

Currently accepted values of C_λ for photon energies up to 35 MeV are tabulated in Reference 4.

There is an obvious problem in the definition of C_λ which lies in the assumption that all the electrons causing ionization in the air cavity arise from the water outside the cavity, i.e., there is no contribution from the air-equivalent walls. In general, there will be a mixture of electrons entering the cavity: some arising in the water and some arising in the wall. The ratio of electrons from these two sources for a given chamber will vary with photon energy. Recently there has been considerable discussion in the literature about the determination of C_λ which indicates that there may be errors of up to 5% in the accepted C_λ values. As yet, there is no generally agreed upon method for the precise calculation of C_λ.

6.6 Applications of Cavity Theory for Electron Fields (C_E)

In principle the absorbed dose in a medium resulting from incident electrons can be determined from a cavity chamber measurement provided the cavity does not perturb the electron field significantly. The dose to a medium from an incident electron beam is given by

$$D_m = D_c \cdot s_c^m \cdot p \tag{6.15}$$

where

D_m is dose to the medium;

D_c is dose to the cavity;

s_c^m is the stopping power ratio for medium to cavity;

p is a perturbation correction caused by introduction of the cavity into the electron field.

The stopping power ratio should be calculated over the spectrum of energies present in the electron field. However, for almost all practical measurements a ratio of stopping powers taken at the mean energy of the electron spectrum will suffice.

The mean energy of an electron beam spectrum at any depth d in an absorbing medium can be determined from[9]

$$E = E_o (1 - d/R_p) \tag{6.16}$$

where

E_o is the initial energy of the beam;

R_p is the practical, or extrapolated, electron range.

In practice an absolute measurement of dose to the cavity is not made. However, a chamber calibration at Co-60 energies can be used. Dose to an air cavity chamber will be given by

$$D_a = 0.87 \, M \cdot N_c \cdot d_c \tag{6.17}$$

where

M is the chamber reading corrected for temperature and pressure;

N_c is the exposure calibration factor;

d_c is a correction for displacement of the medium by the chamber, as before.

Dose to the medium is determined from Equation 6.15

$$D_m = 0.87 \, M \cdot N_c \cdot d_c \cdot s_a^m \, p$$

In a manner analogous to the derivation of C_λ, a factor C_E is defined for use in dose measurements in electron beams:[10]

$$C_E = 0.87 \, d_c \cdot p \cdot s_a^m \tag{6.18}$$

Perturbation factors for thimble chambers in water are given in Reference 10. When a thin parallel plate chamber is used, p can be assumed to be unity.

Values of C_E calculated according to Equation 6.18 and taking into consideration the electron spectrum are less uncertain than values of C_λ, especially for thin window parallel plate chambers since electrons causing ionization in the chamber will enter the chamber undisturbed from the medium outside the chamber. The major uncertainty in the use of C_E is in the determination of the electron spectrum (or mean energy) at the point of measurement.

6.7 <u>Average Energy Associated with the Formation of One Ion Pair (W)</u>

To determine the absorbed dose in a medium using a gas cavity it is necessary to determine the absorbed dose in the gas. Since ionization in the gas is generally the quantity measured we must know the amount of energy deposited in the gas in the production of ionization. The amount of energy lost by an electron by all processes averaged over the entire electron track for each ion pair formed is denoted by W. The best experimental determination of W for air to date have yielded a value of 33.7 eV/ion pair for electrons of energy greater than 20 keV.[8] Below 20 keV, W is expected to be somewhat energy dependent but can be assumed to be constant for energies greater than 20 keV. The value of W is greater than the actual ionization potential of the gas because some energy is lost in processes other than ionization, such as excitation. Values of W for other gases and particles other than electrons are tabulated in ART and NBS Handbook 85.[8]

The value of W for gas mixtures can be calculated from the relationship

$$\frac{1}{W} = \sum_i \left(\frac{P_i}{W_i}\right)$$

where P_i are the relative partial pressures of the gases.

REFERENCES

1. L. H. Gray, Proc. Royal Soc. $\underline{A122}$, 647 (1929).
2. L. H. Gray, Proc. Royal Soc. $\underline{A156}$, 578 (1936).
3. NCRP Report 27, <u>Stopping Powers for Use with Cavity Chambers</u>, National Committee on Radiation Protection and Measurements, Natl. Bur. Std. (U.S.) Handbook 79 (1961).
4. ICRU Report 14, <u>Radiation Dosimetry: X-Rays and Gamma Rays with Maximum Photon Energies Between 0.6 and 50 MeV</u>, International Commission on Radiation Units and Measurements (1969).
5. T. E. Burlin and F. K. Chan, The Influence of Interfaces on Dosimeter Response, <u>Proceedings of the Symposium on Microdosimetry, Ispra (Italy), November 13-15, 1967</u> (European Communities, Brussels, 1968).
6. F. H. Attix, Health Physics $\underline{15}$, 49 (1968).
7. F. K. Chan and T. E. Burlin, Health Physics $\underline{18}$, 325 (1970).
8. ICRU Report 10b, <u>Physical Aspects of Irradiation,</u> International Commission on Radiation Units and Measurements, Natl. Bur. Std. (U.S.) Handbook 85 (1964).
9. D. Harder, Energiespektren schneller Electronen in verschiedenen Tiefen, Symposium on High Energy Electrons--Montreux, (1964) A. Zuppinger and G. Poretti (eds.) (Springer-Verlag, Berlin, 1965).
10. ICRU Report 21 <u>Radiation Dosimetry: Electrons with Initial Energies Between 1 and 50 MeV</u>, International Commission on Radiation Units and Measurements, Washington, D.C. (1969).

MAIN REFERENCES

(ART)   F. H. Attix, W. C. Roesch, and E. Tochilin (eds.), <u>Radiation Dosimetry</u>, Second Edition, Volume I, Fundamentals (Academic

Press, New York, 1968).

(MT) K. Z. Morgan and J. E. Turner (eds.), <u>Principles of Radiation Protection</u> (Krieger Publishing Co., New York, 1973).

APPENDIX

The appendix contains graphs of functions useful in making flux density and dose calculations for various source geometries as discussed in Chapter 5. Figures A.1 through A.13 show the exponential integrals E_1 and E_2 along with e^{-x}. Figures A.14 through A.19 graph the Sievert integrals (F functions). The graphs in Figs. A.20 through A.24 show the parameters necessary for determining self-absorption in cylindrical and spherical sources. Figures A.25 through A.30 show buildup factors in lead, iron and water. The parameters plotted in Figs. A.31 through A.36 are required for calculating buildup factors in iron, water, lead and concrete.

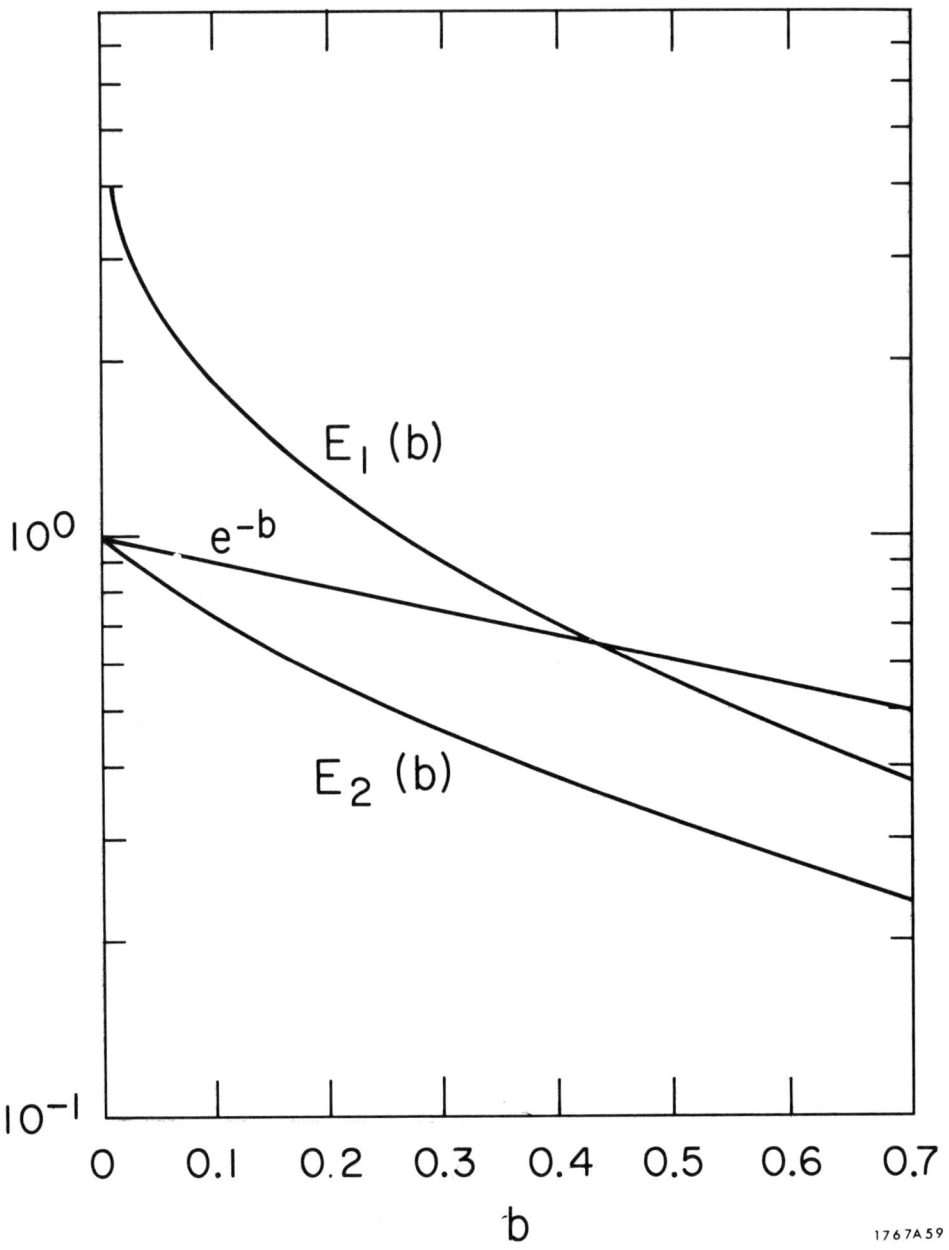

FIG. A.1

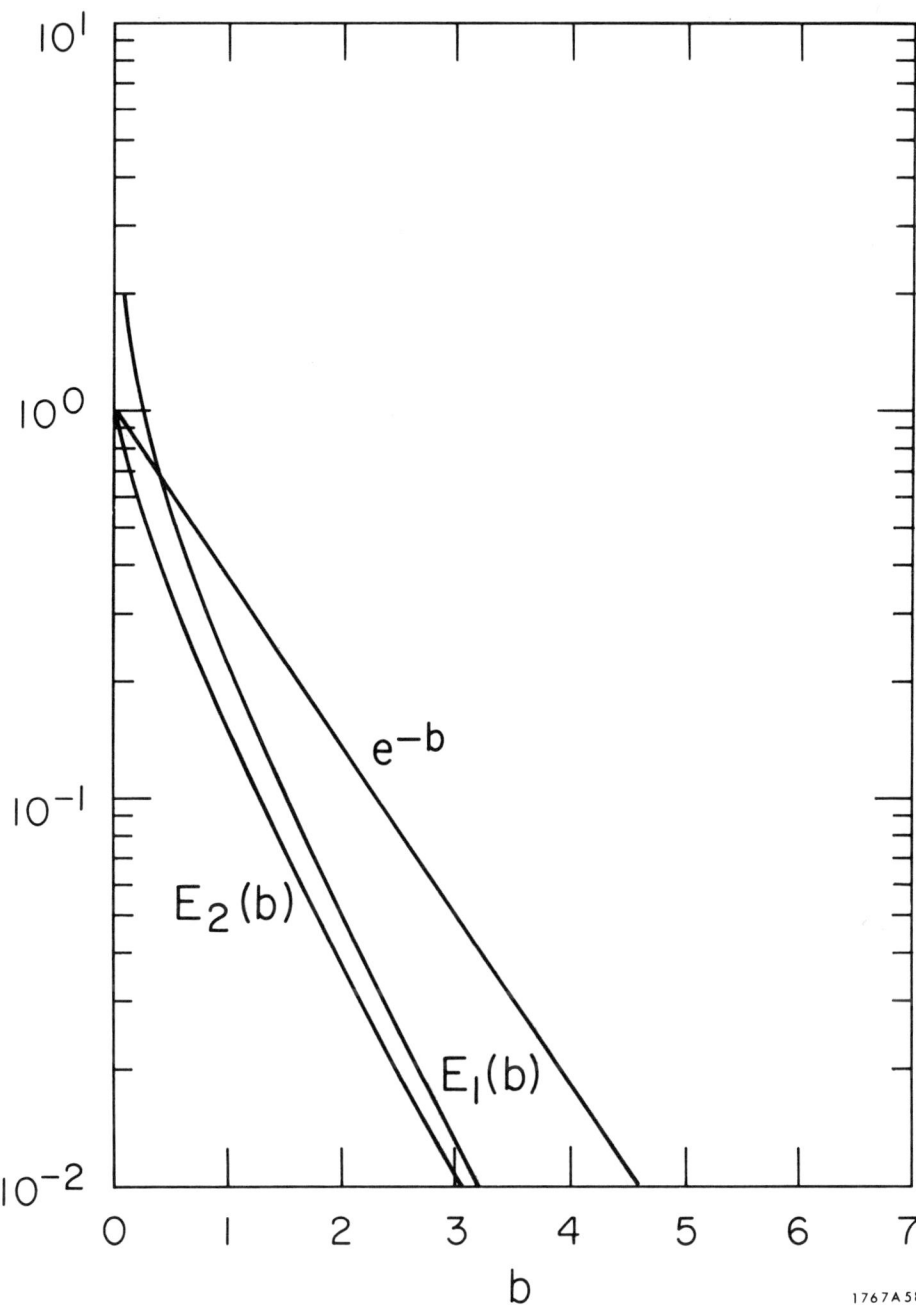

FIG. A.2

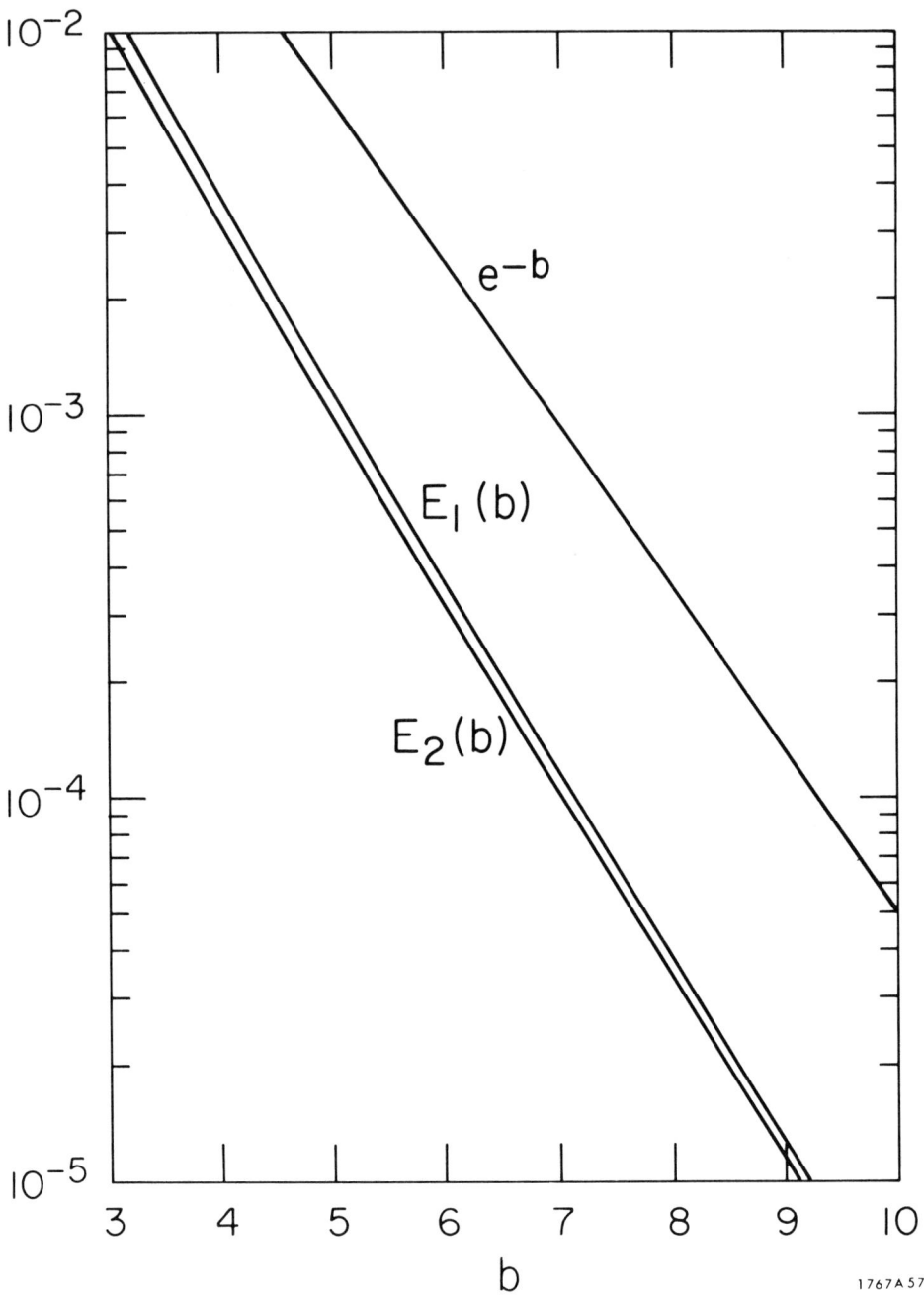

FIG. A.3

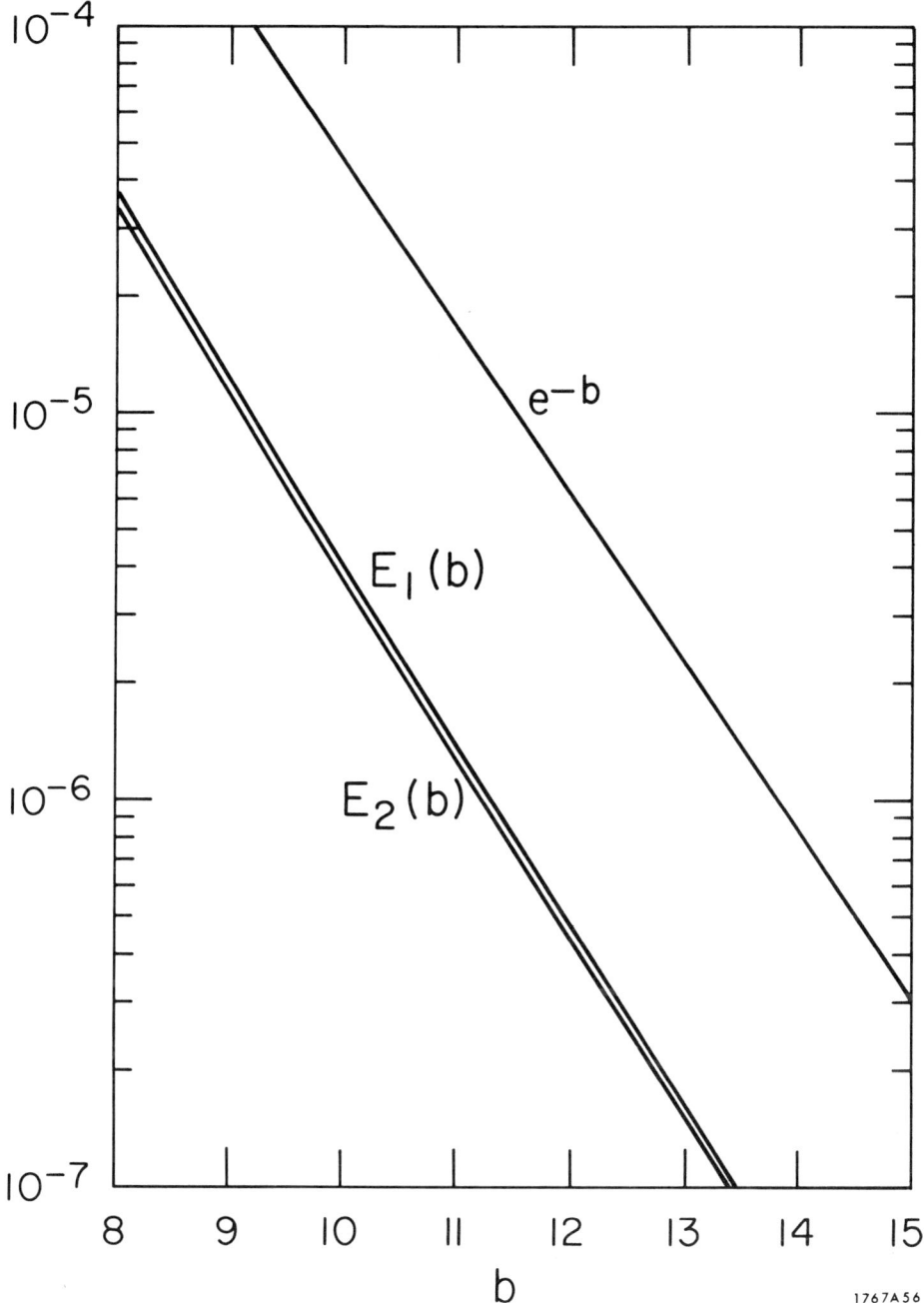

FIG. A.4

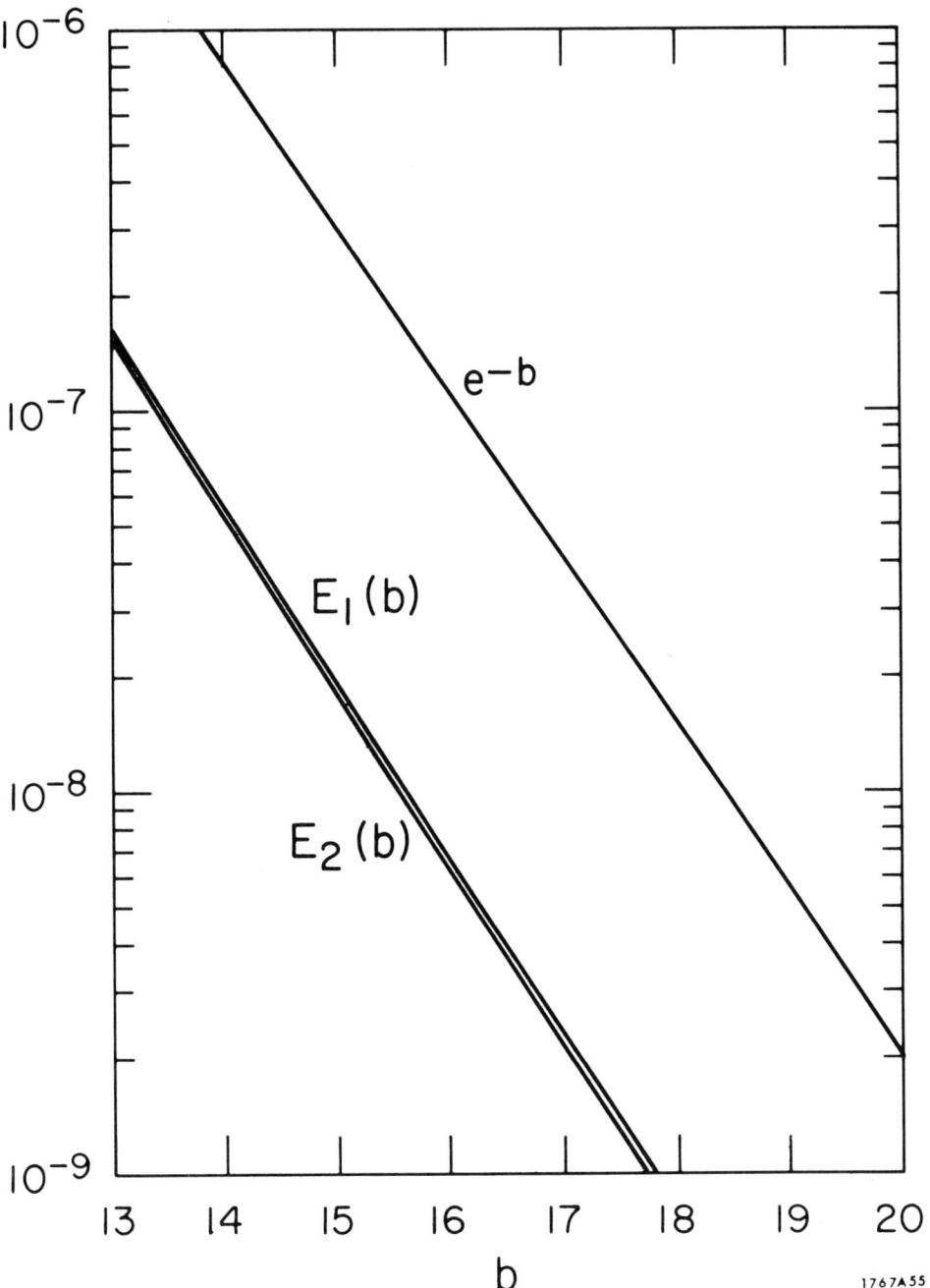

FIG. A.5

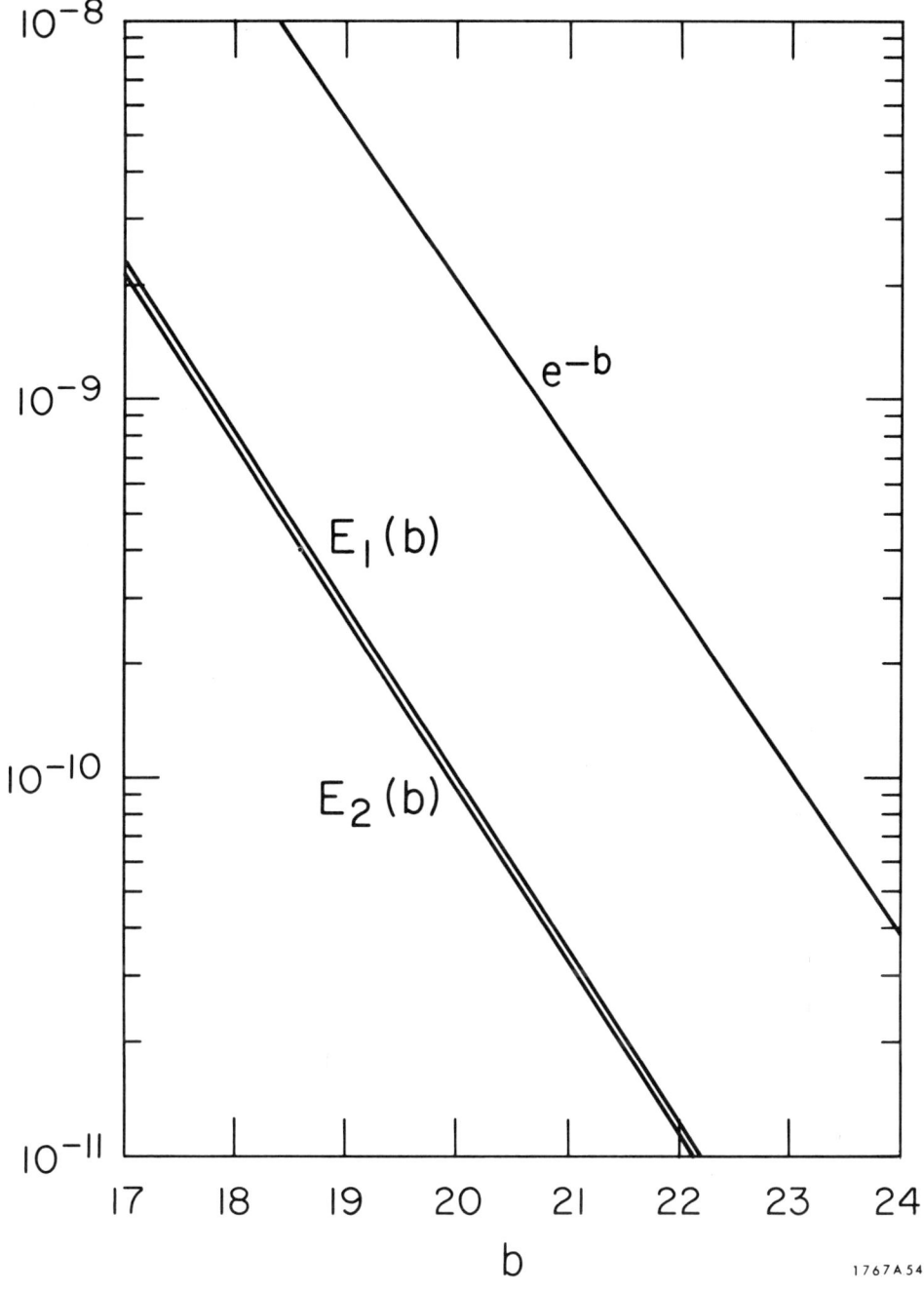

FIG. A.6

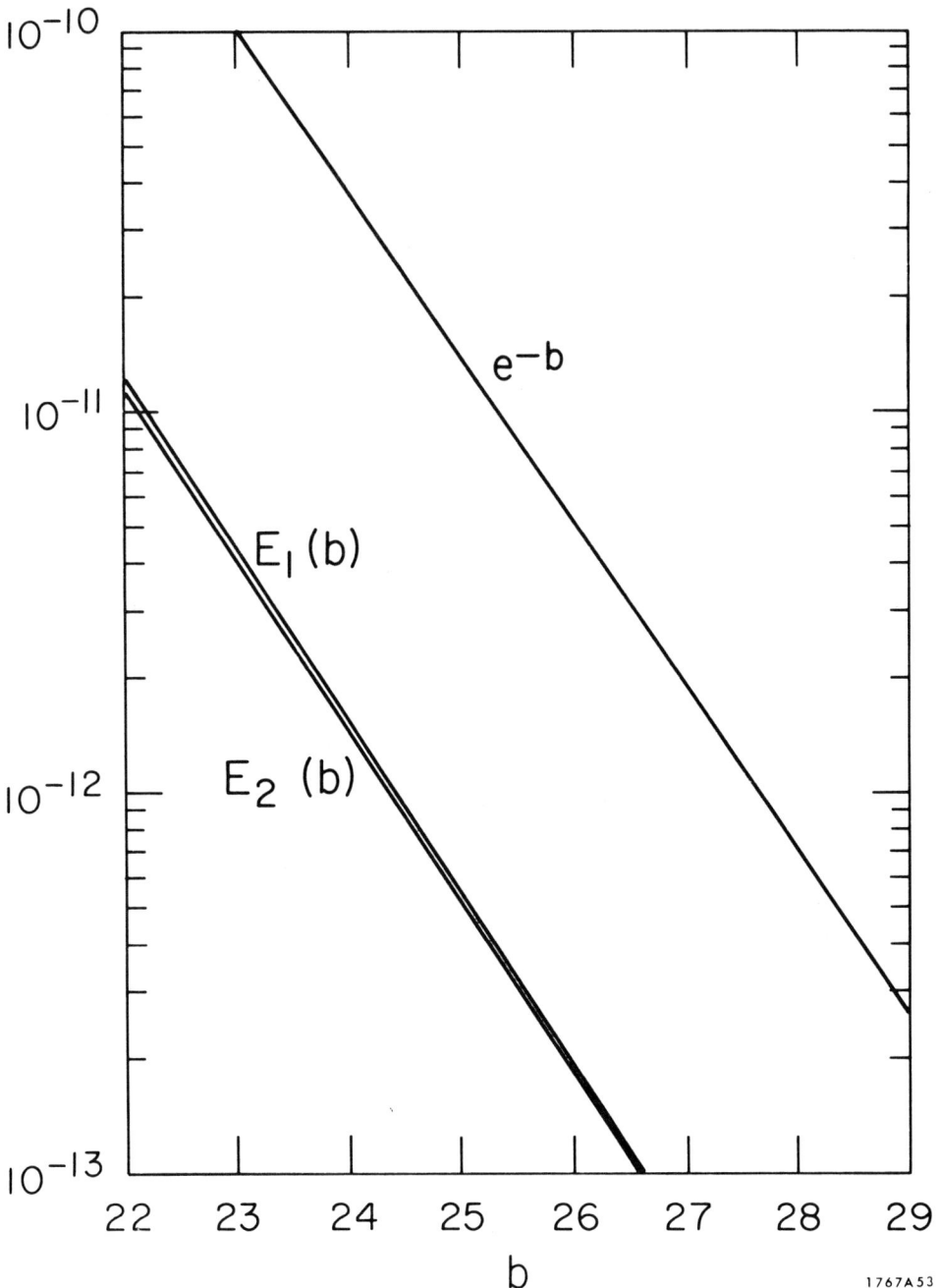

FIG. A.7

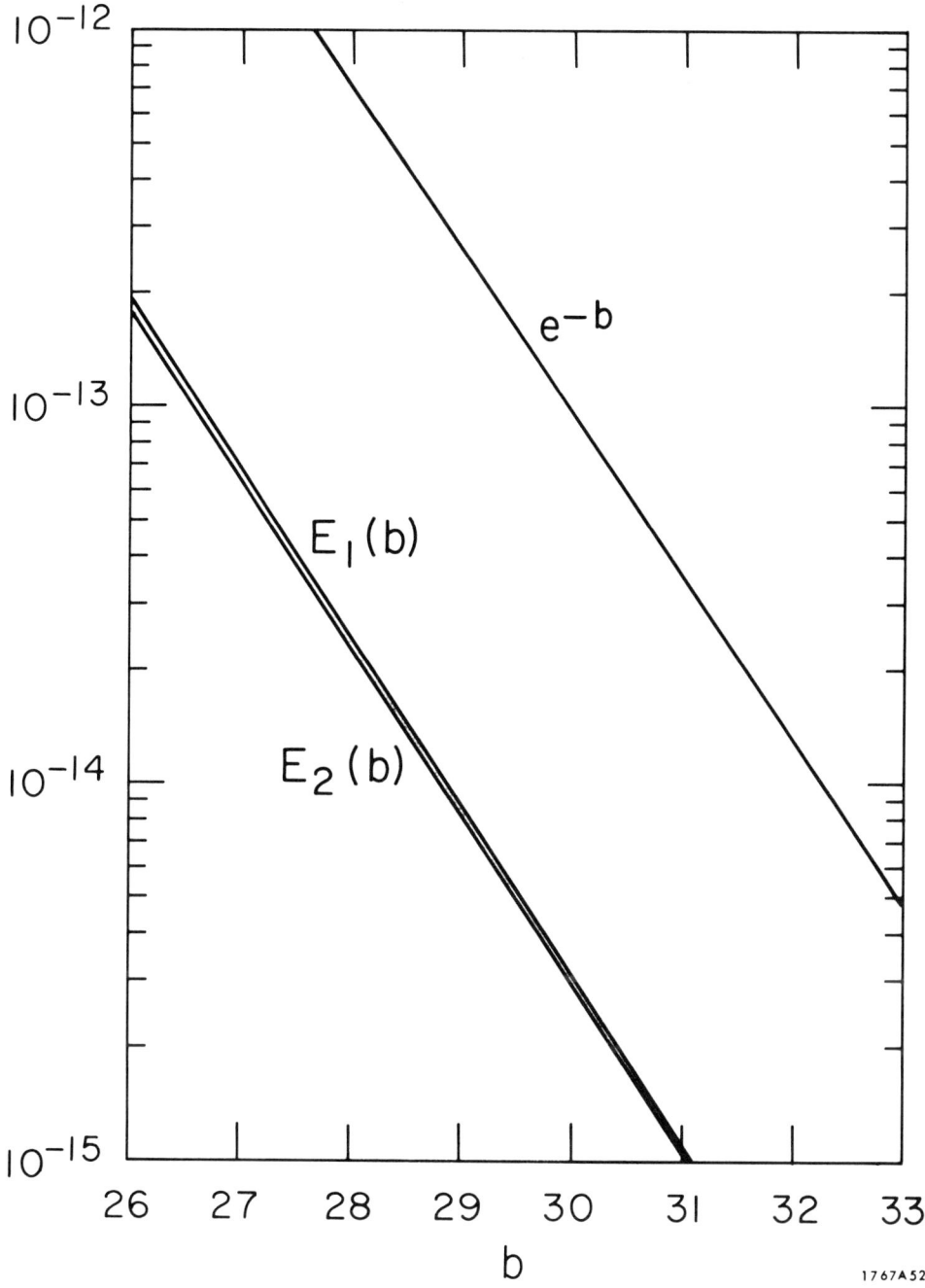

FIG. A.8

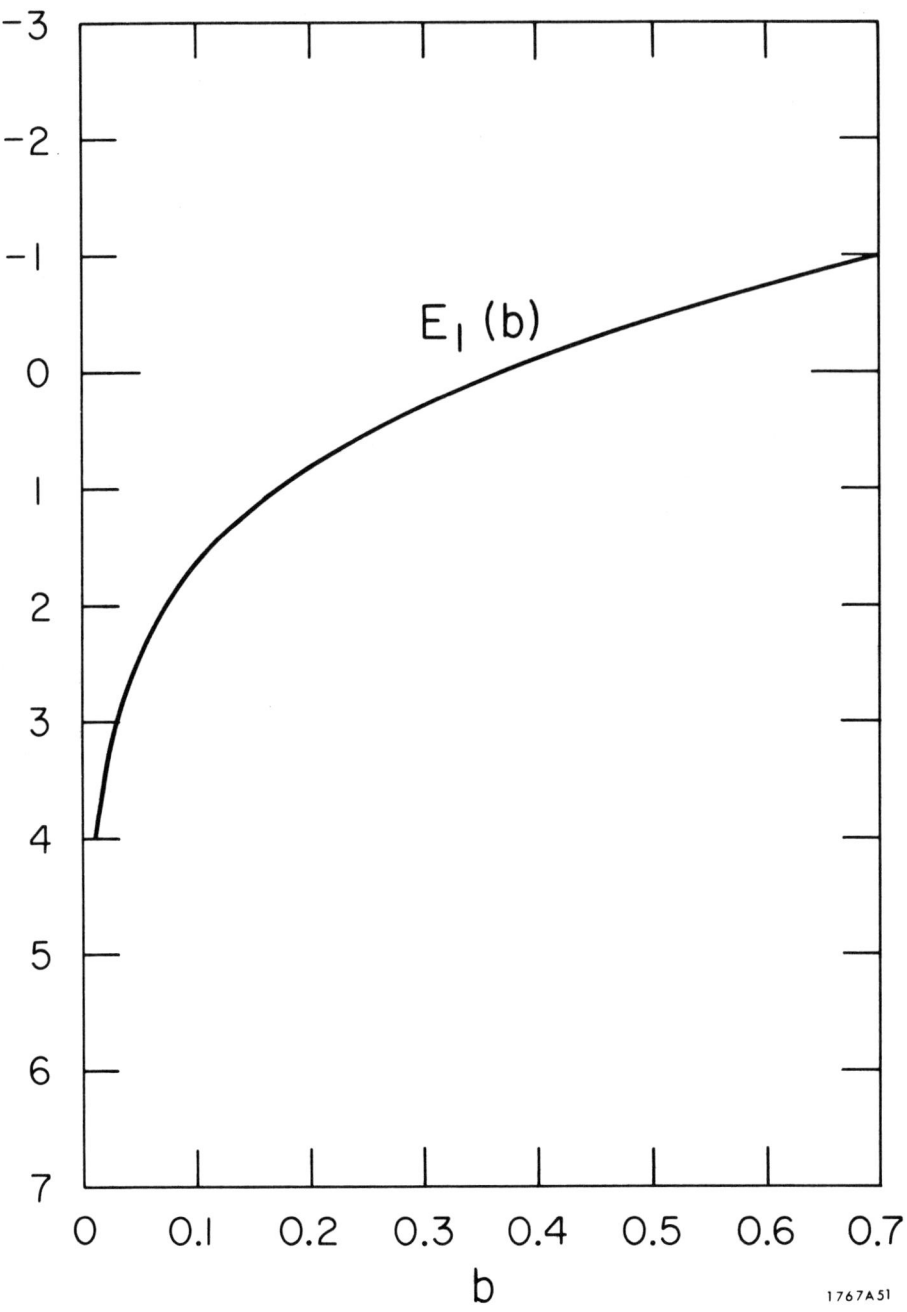

FIG. A.9

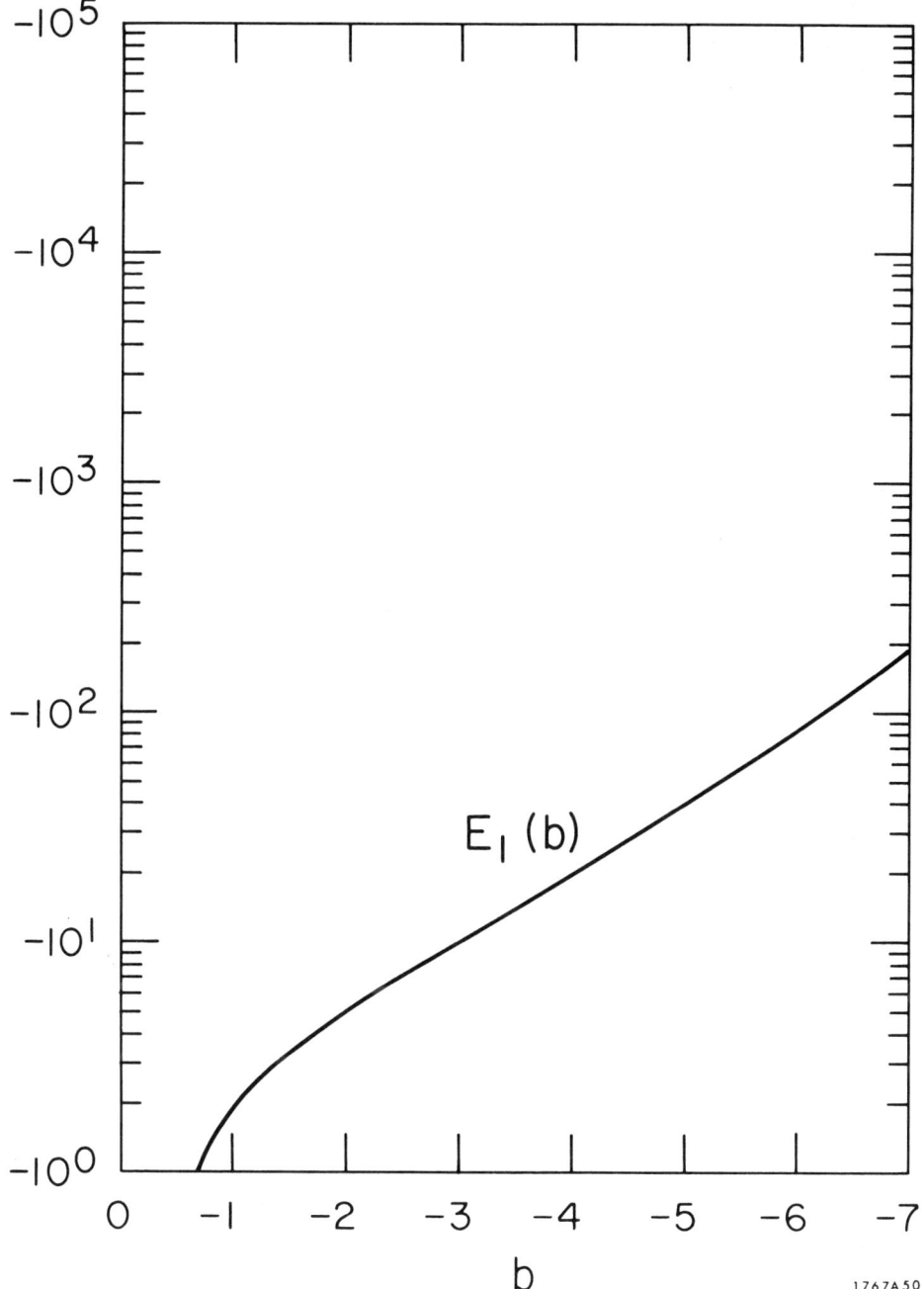

FIG. A.10

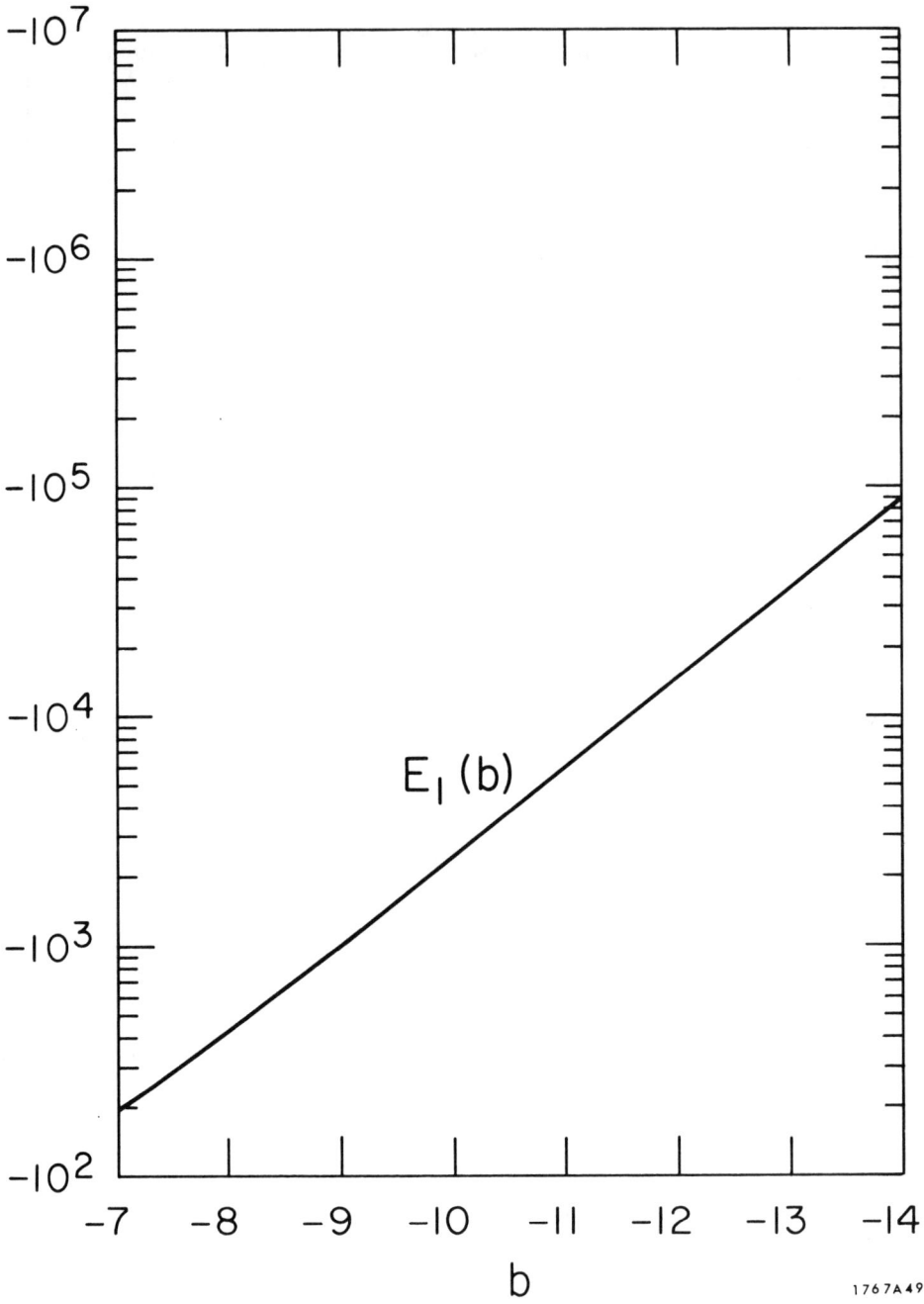

FIG. A.11

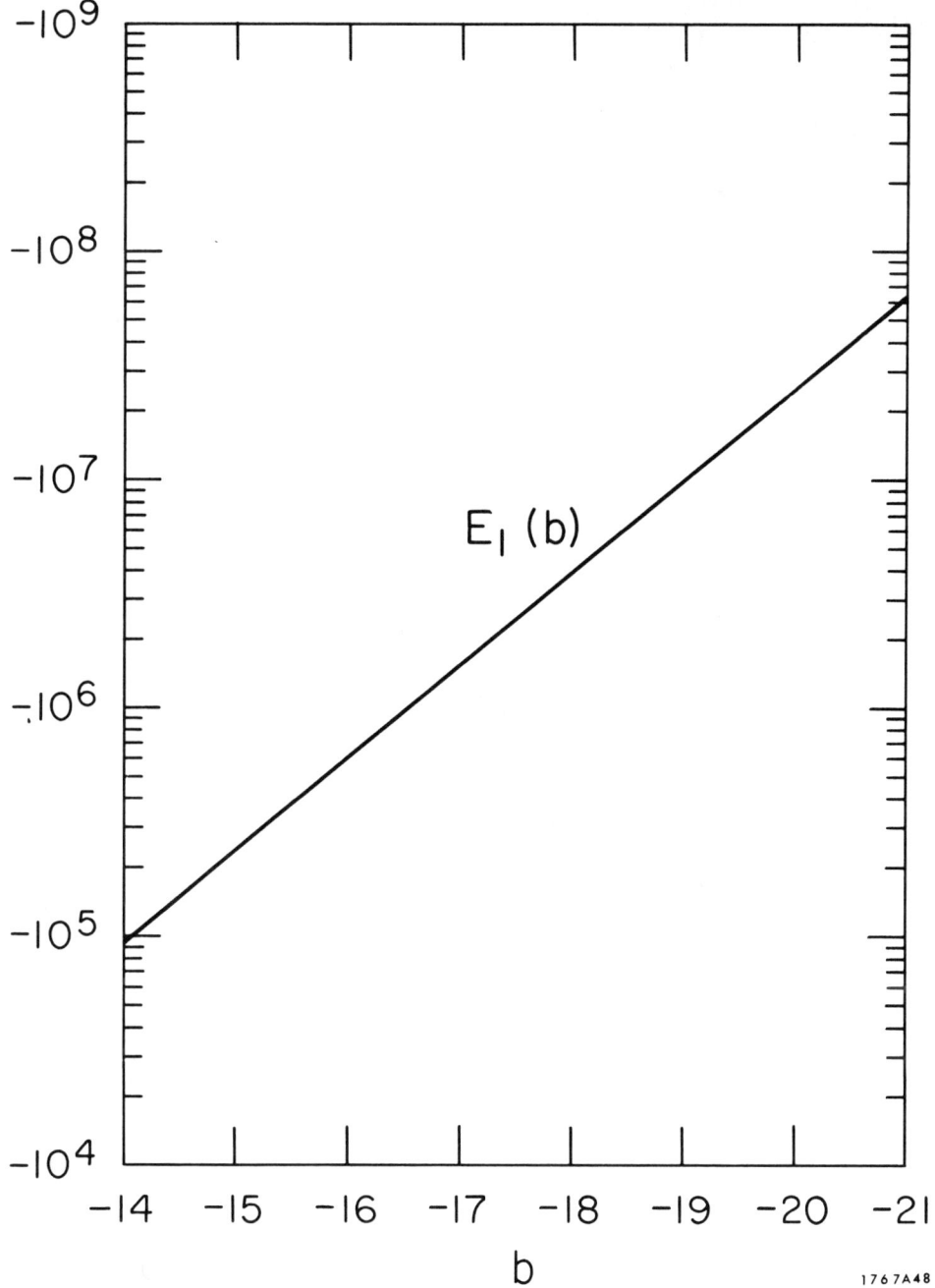

FIG. A.12

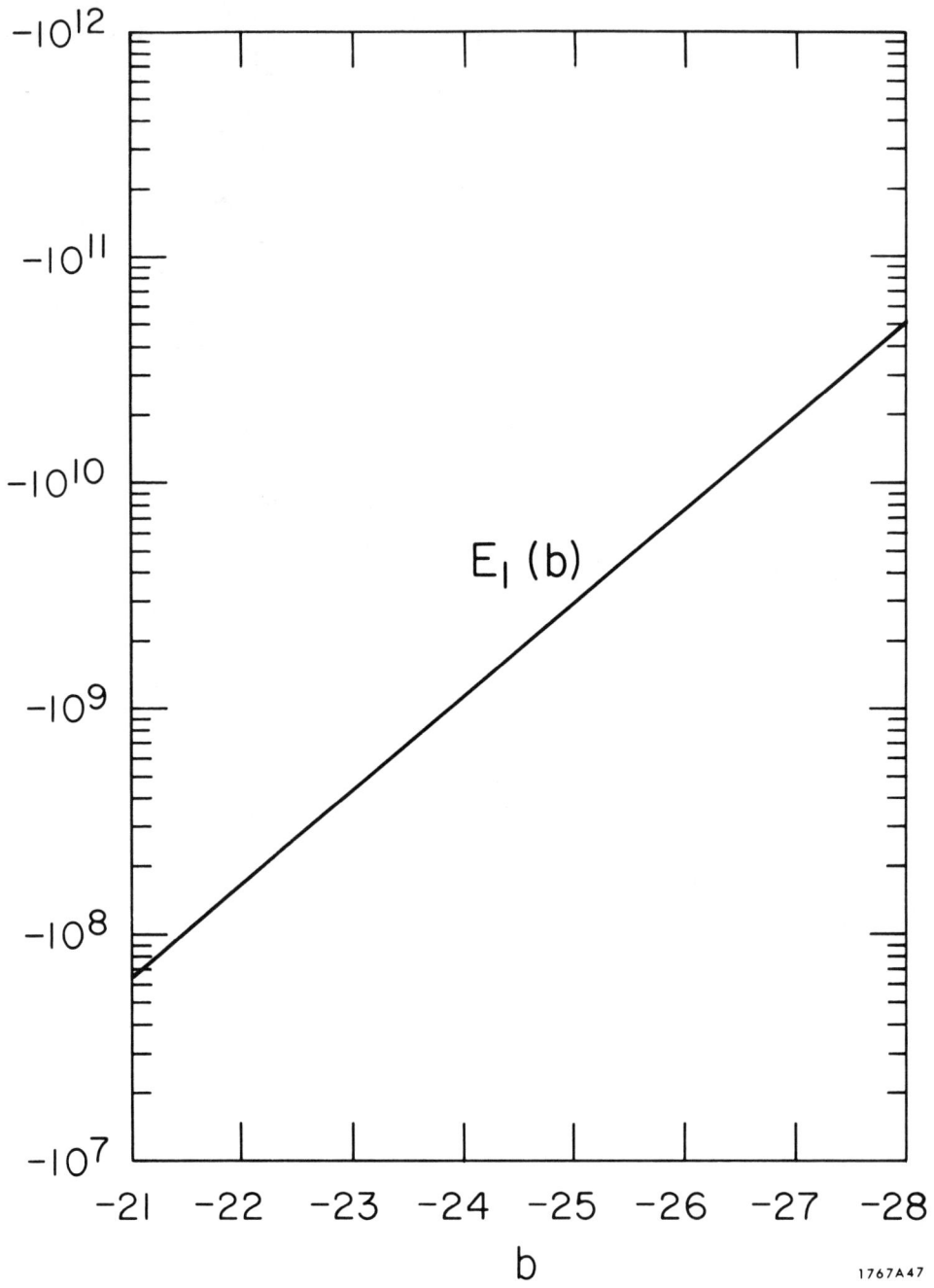

FIG. A.13

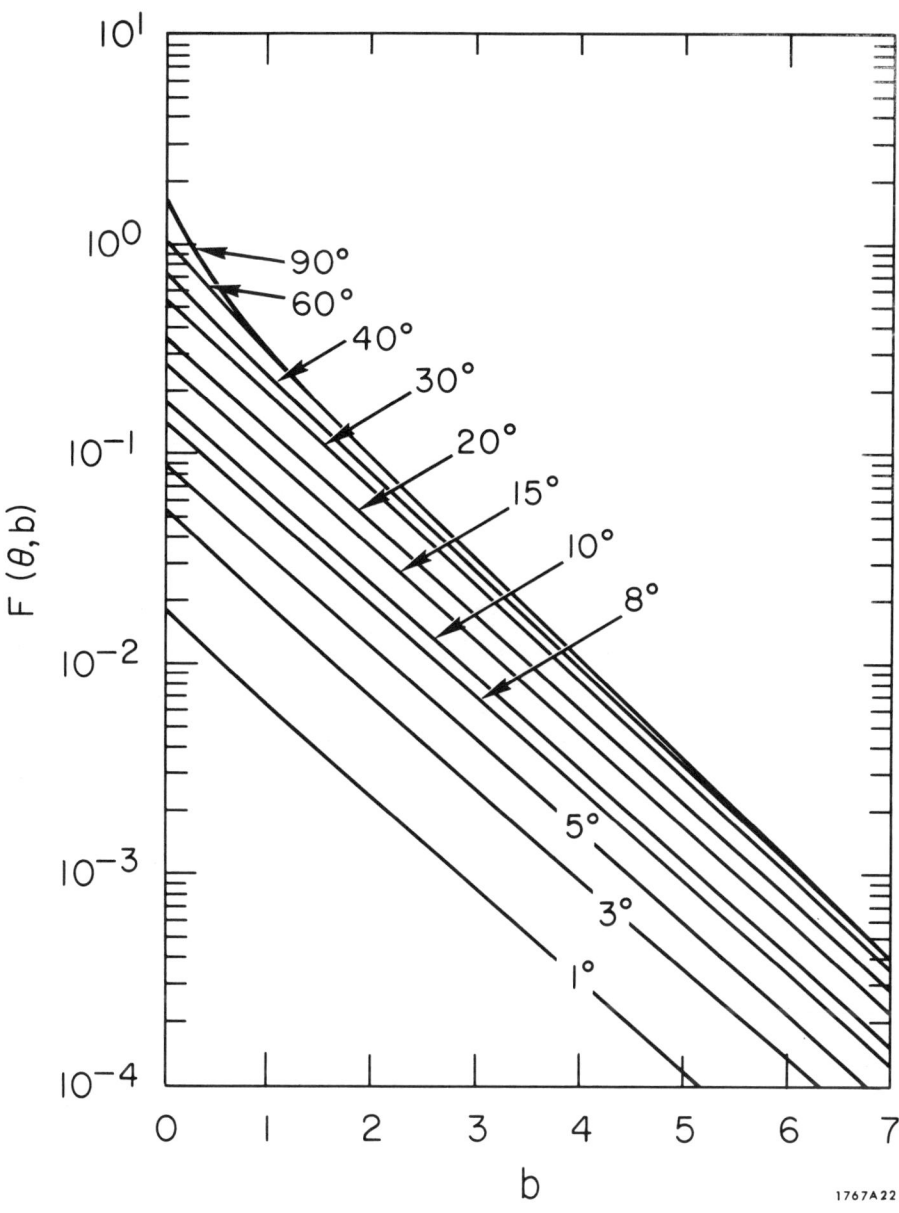

FIG. A.14

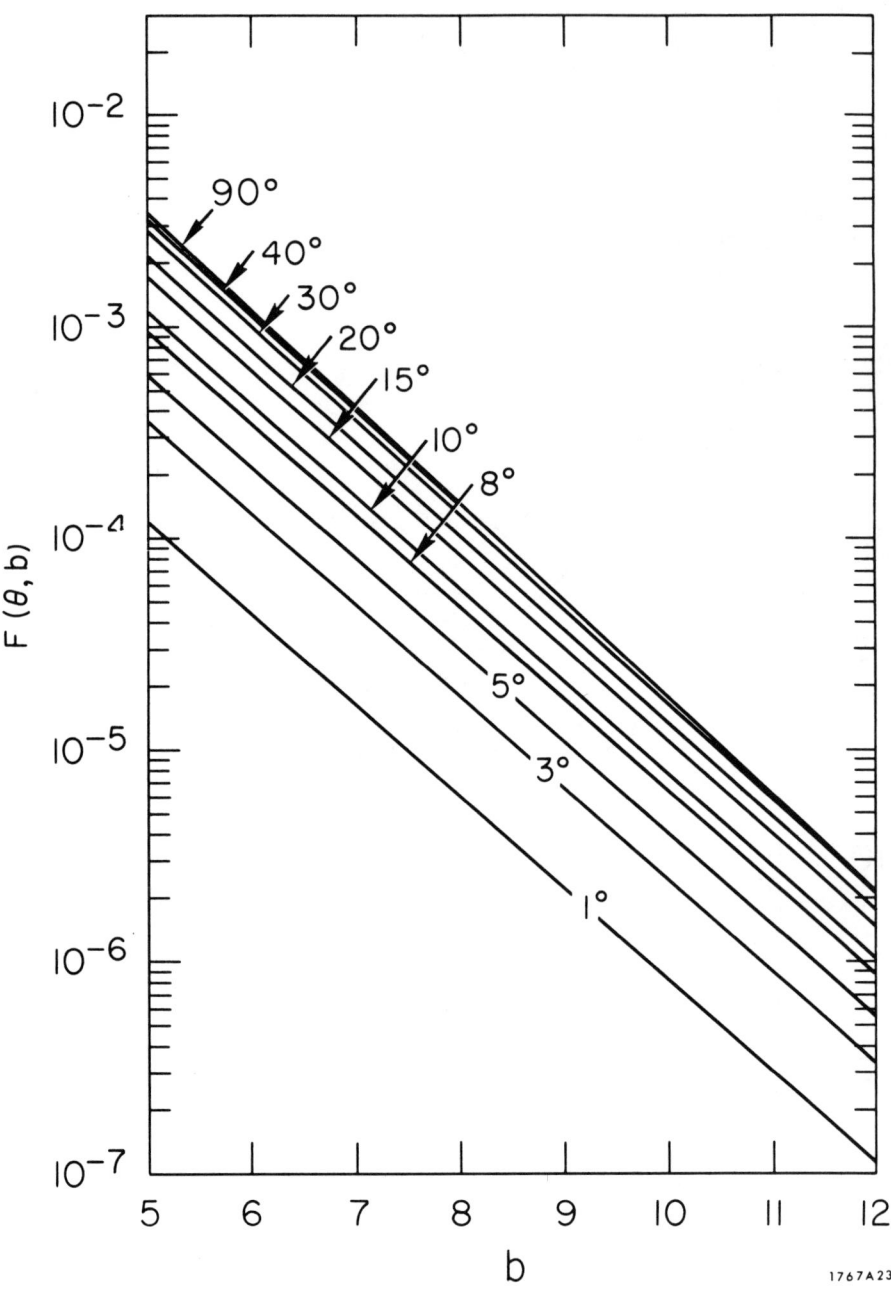

FIG. A.15

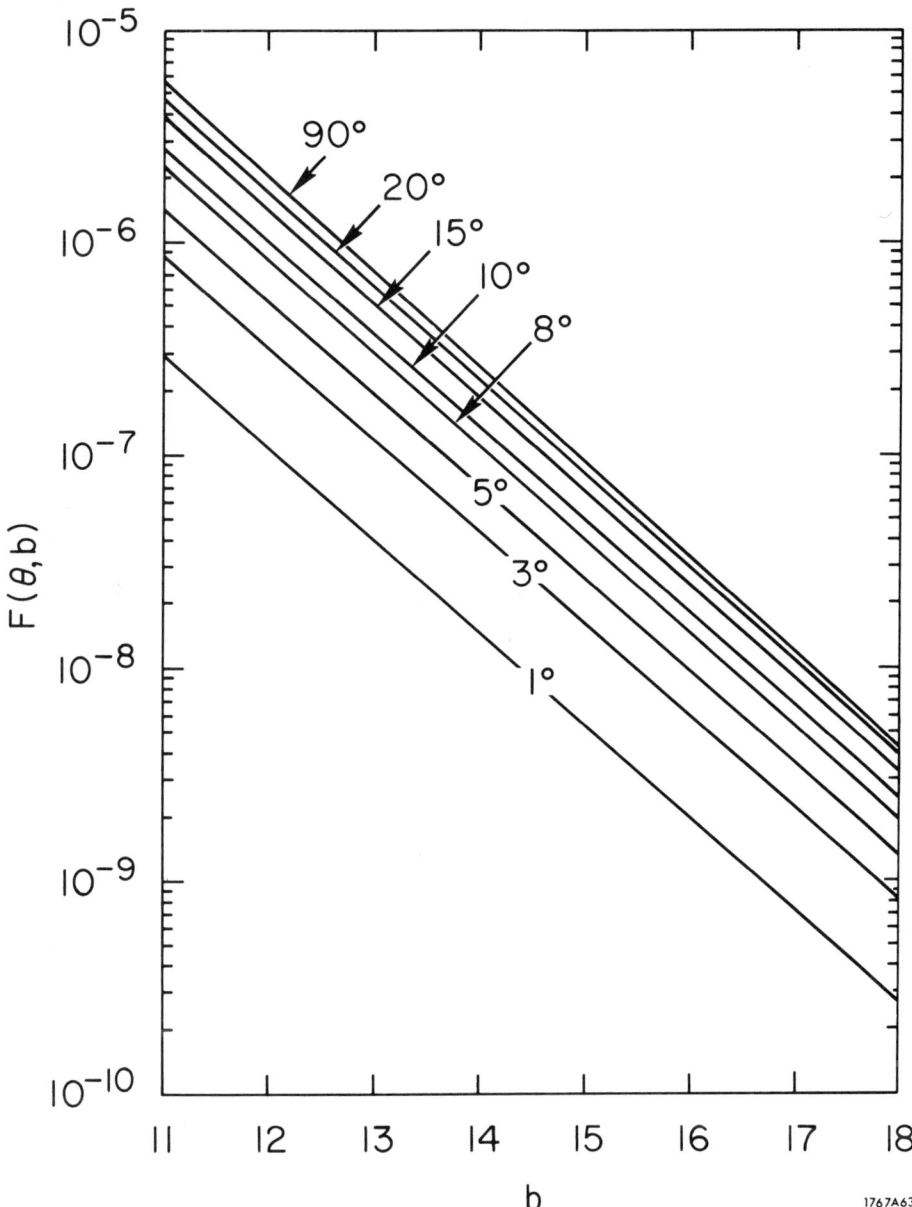

FIG. A.16

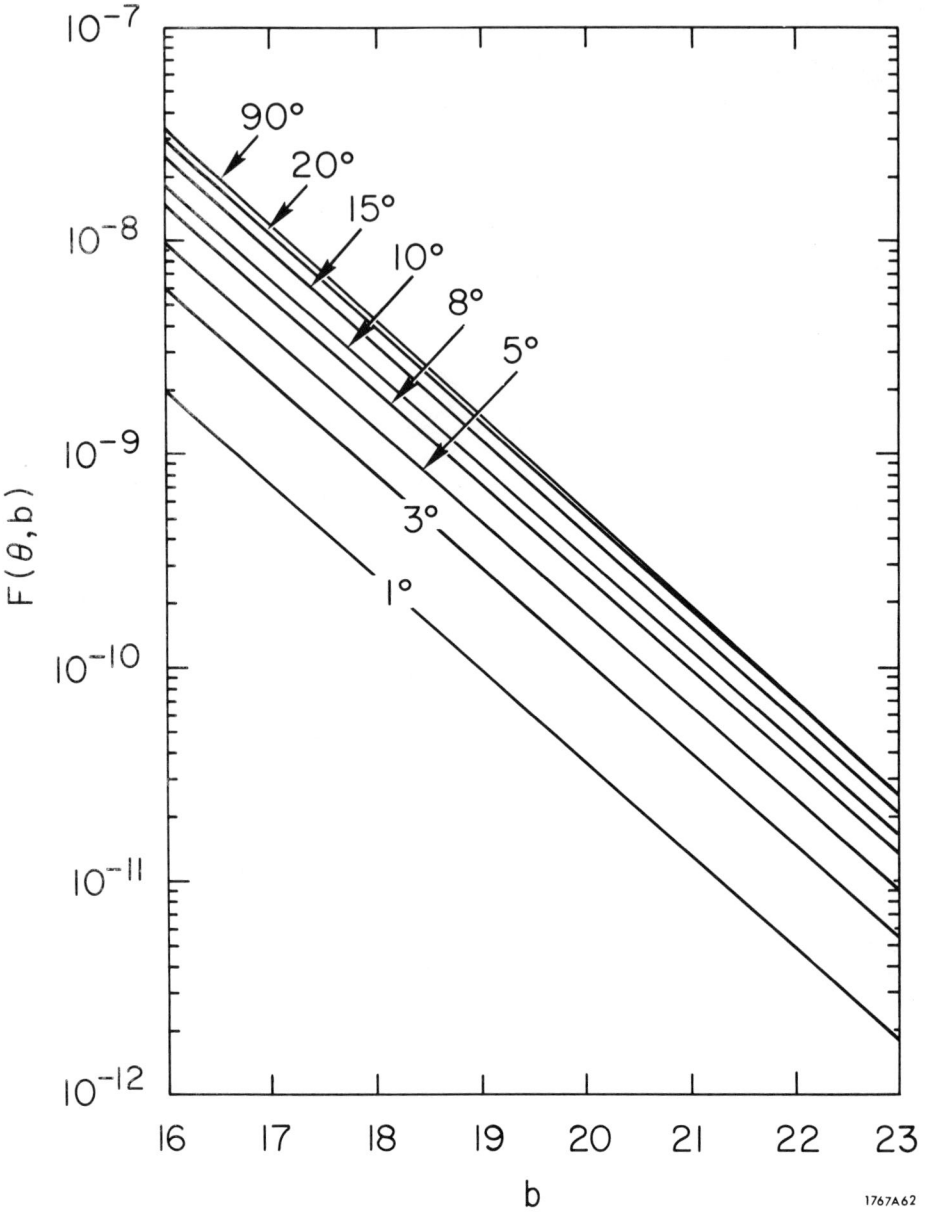

FIG. A.17

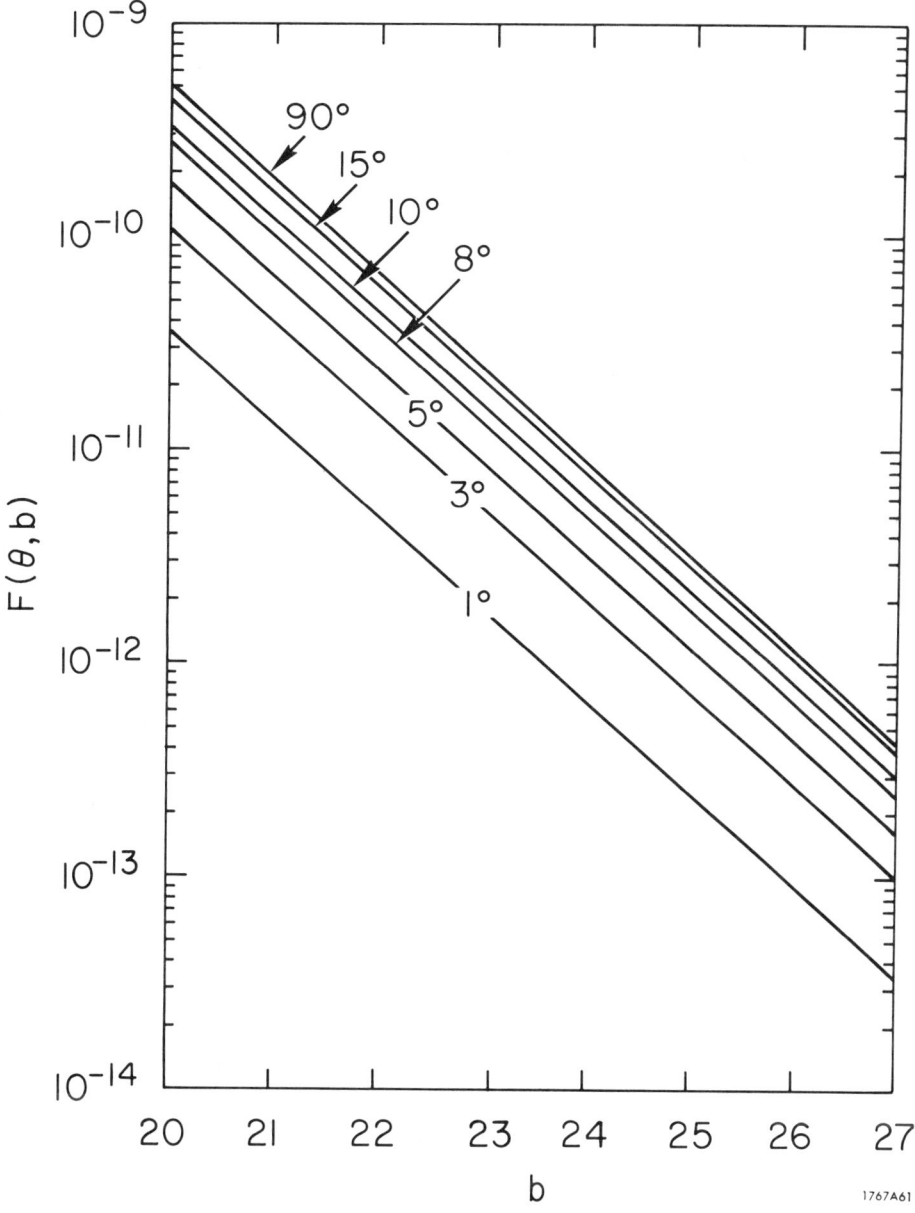

FIG. A.18

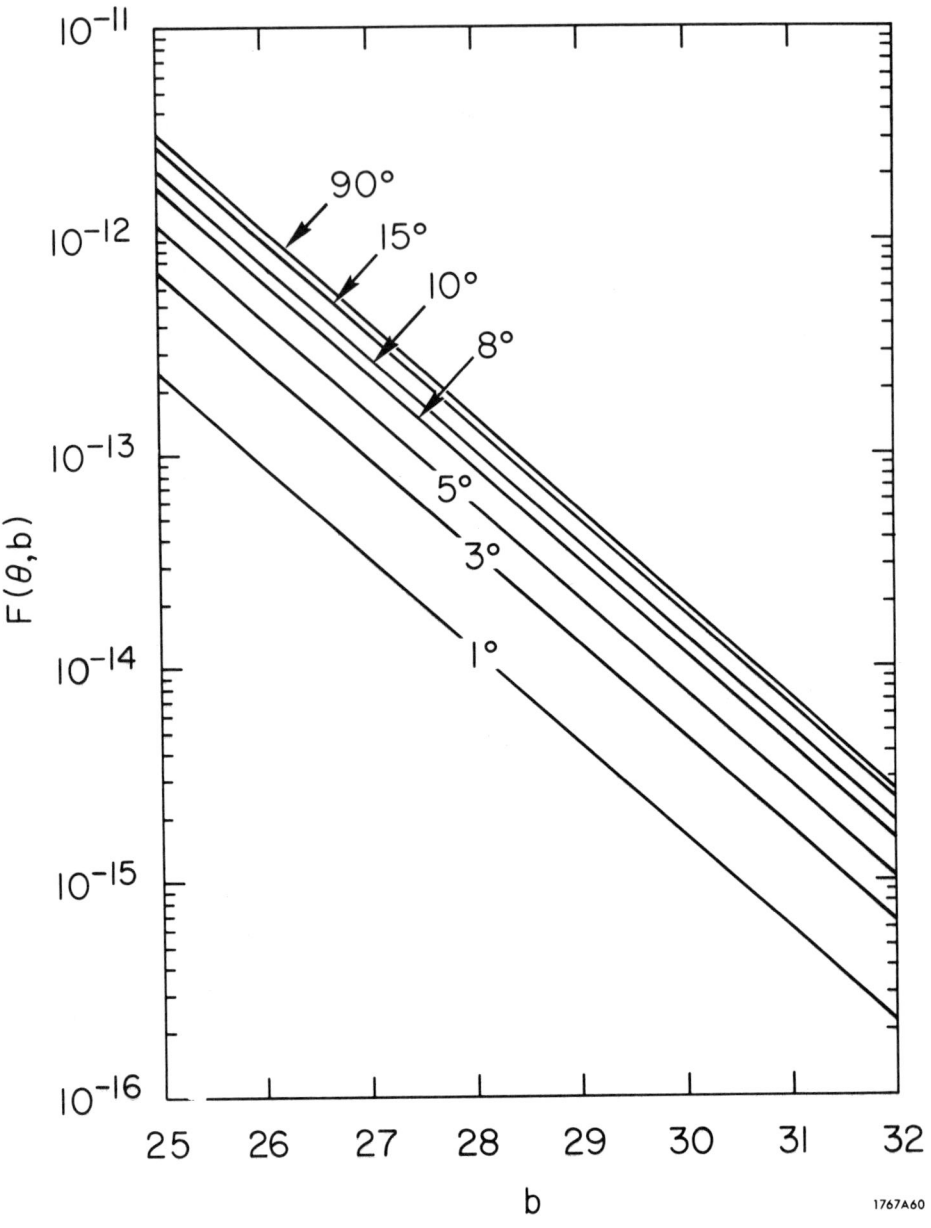

FIG. A.19

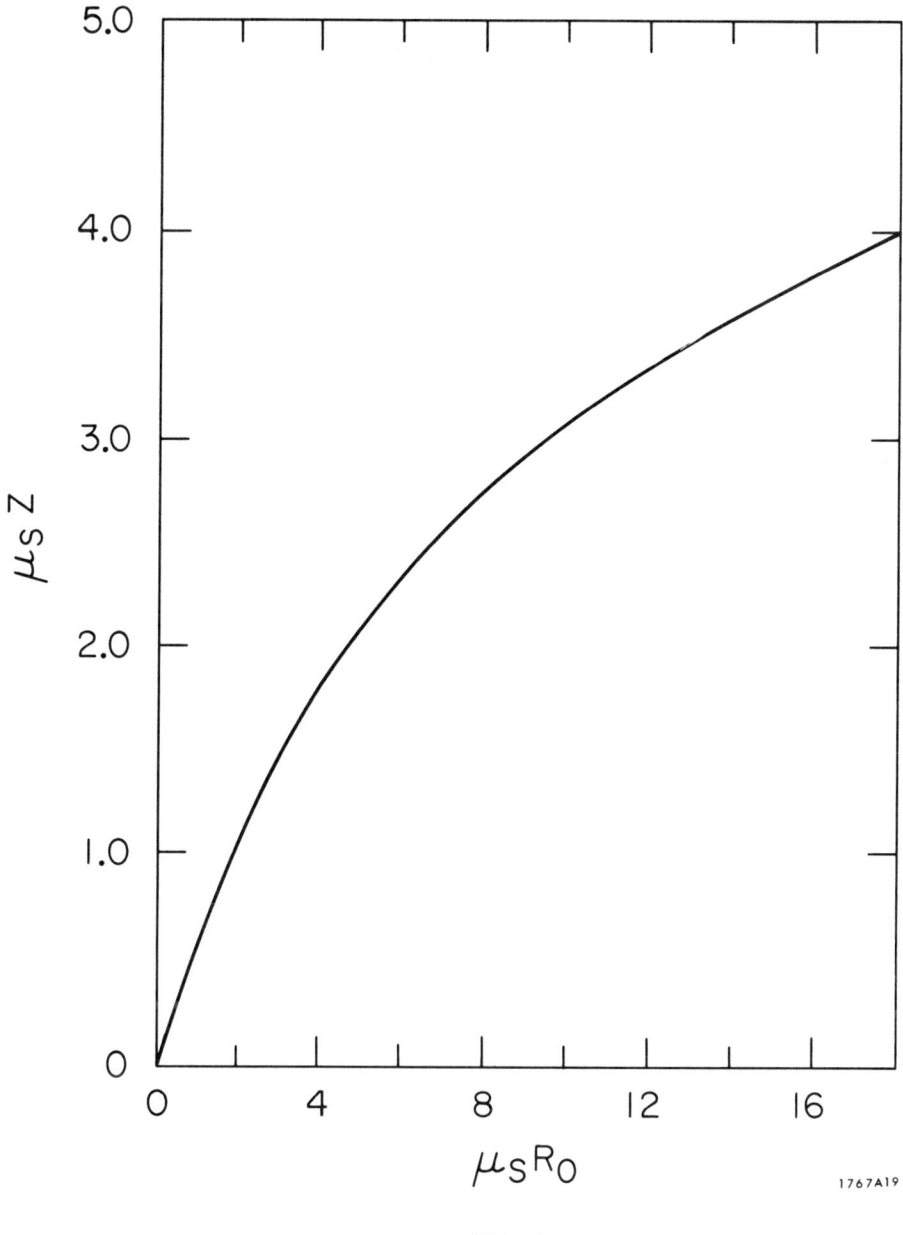

FIG. A.20

Self-absorption distance, Z, of a cylinder for $a/R_0 \geq 10$.

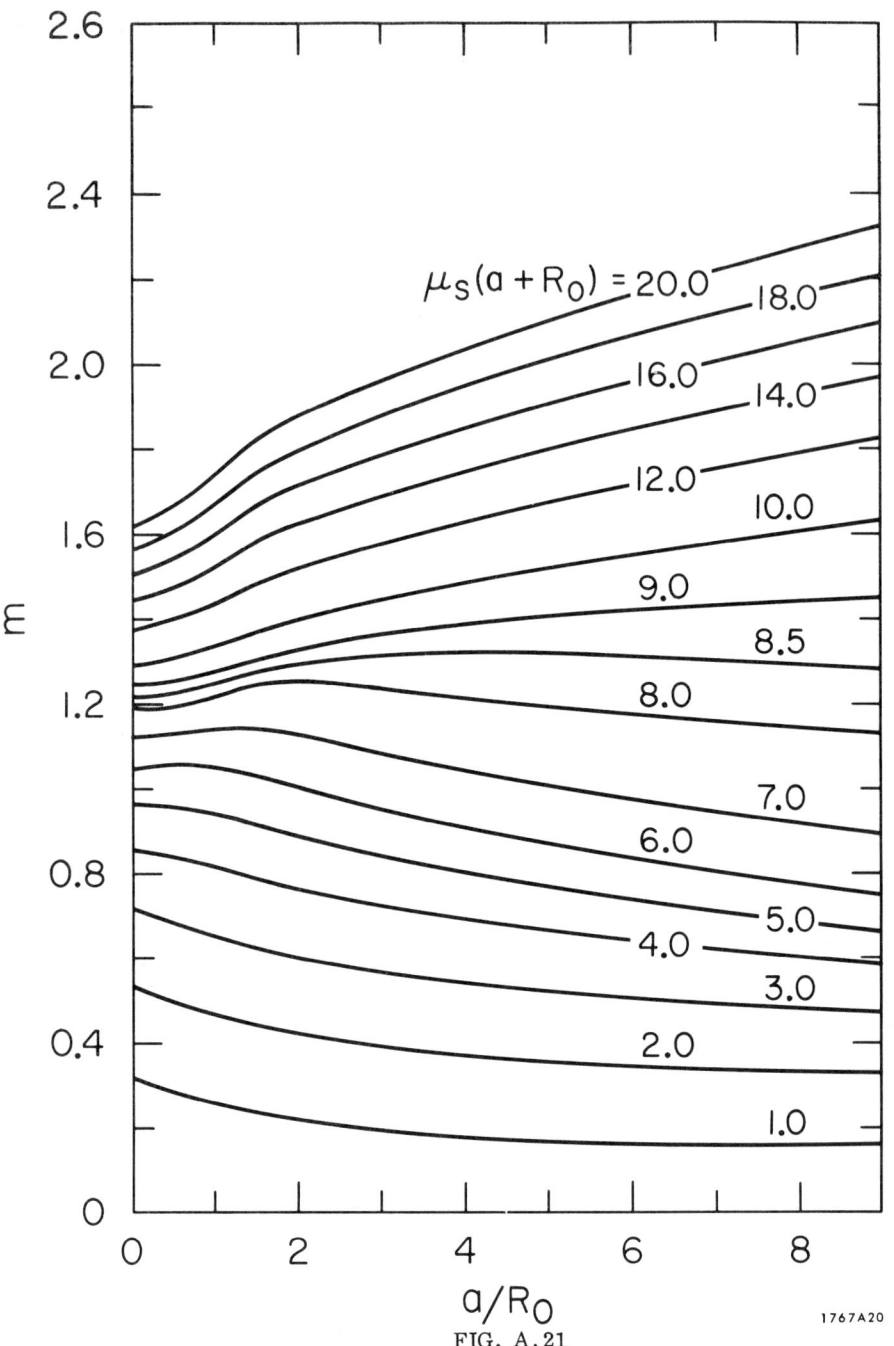

FIG. A.21

Self-absorption distance, Z, of a cylinder for $a/R_0 < 10$.

Note: Use in conjunction with Fig. A.22

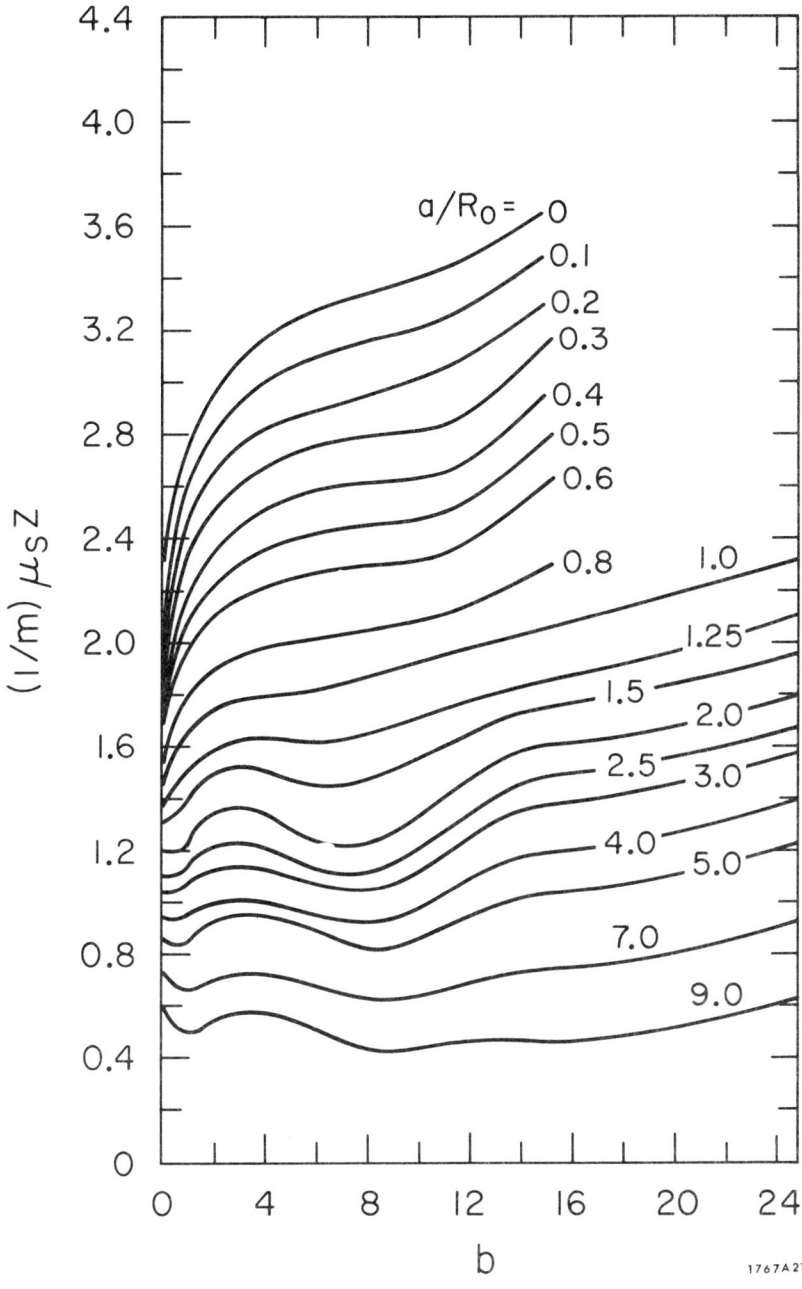

FIG. A.22

Self-absorption distance, Z, of a cylinder for $a/R_0 < 10$.

Note: Use in conjunction with Fig. A.21

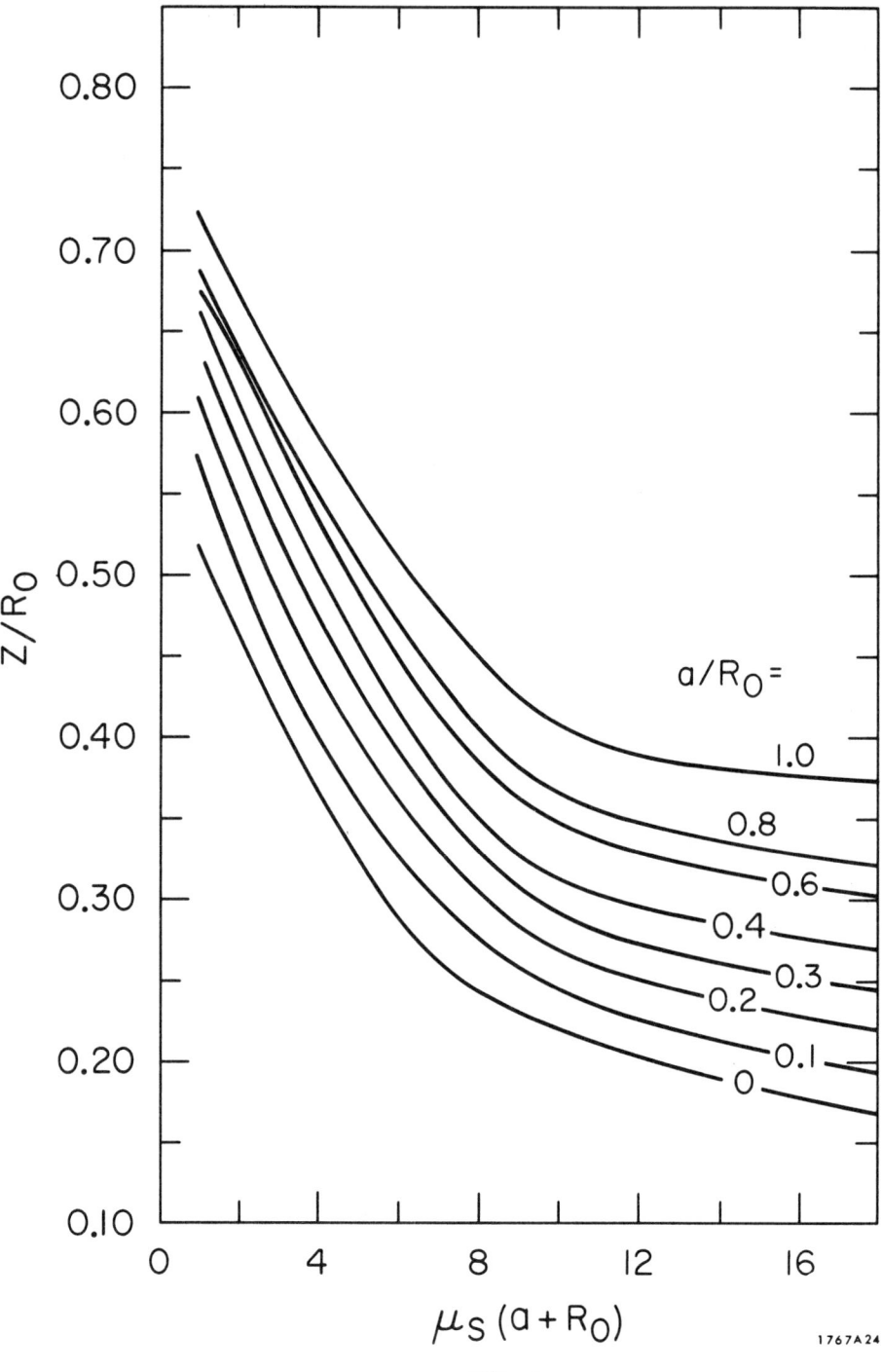

FIG. A.23

Self-absorption distance, Z, of a sphere for $a/R_0 < 1$.

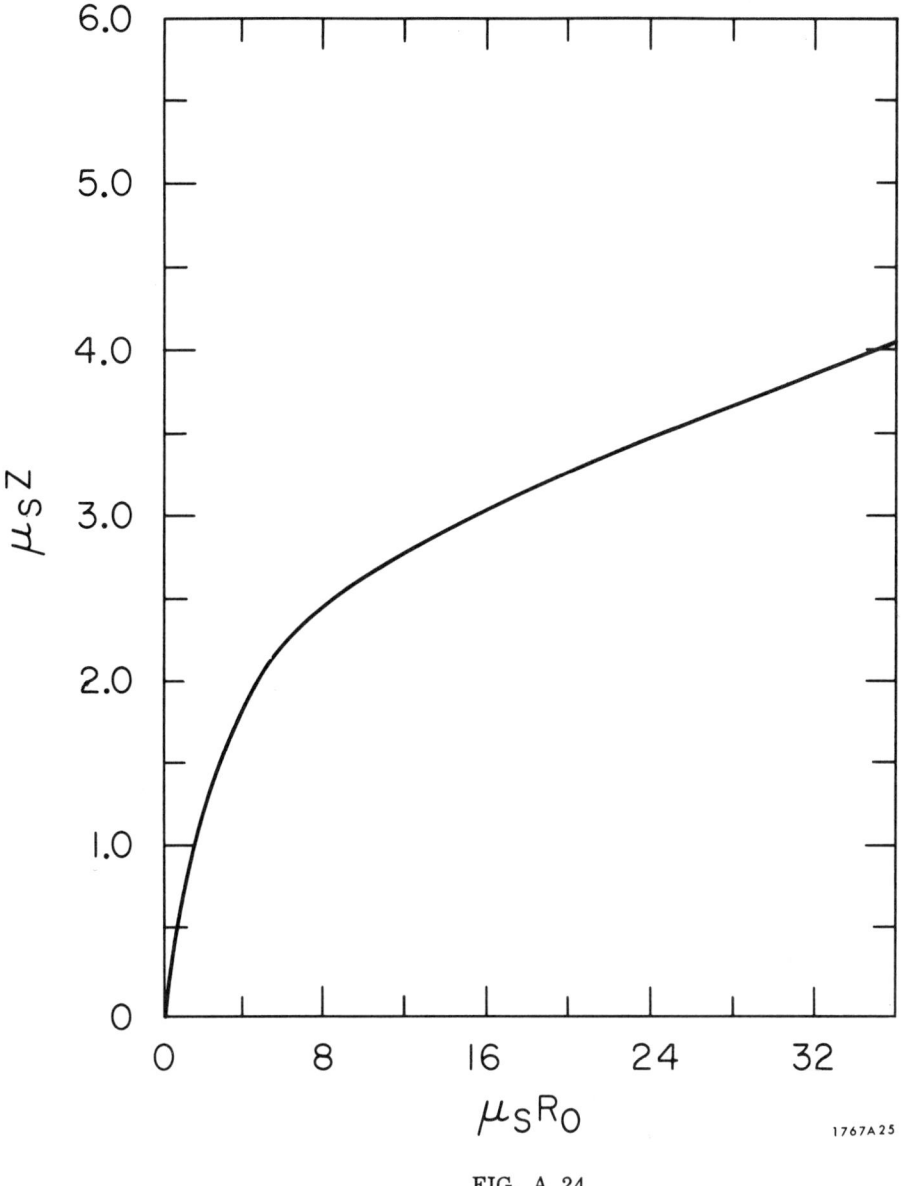

FIG. A.24

Self-absorption distance, Z, of a sphere for $a/R_0 \geq 1$.

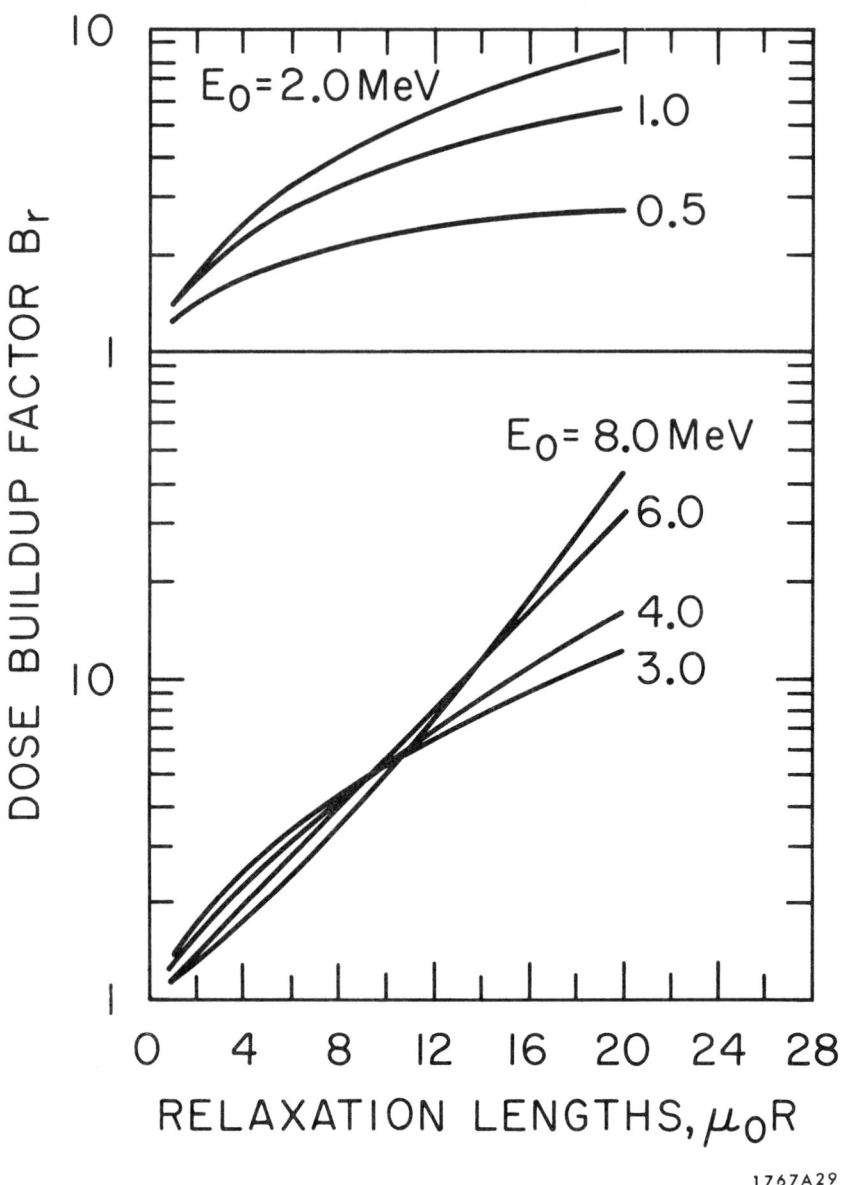

FIG. A.25

Dose buildup factor in lead for a point isotropic source.

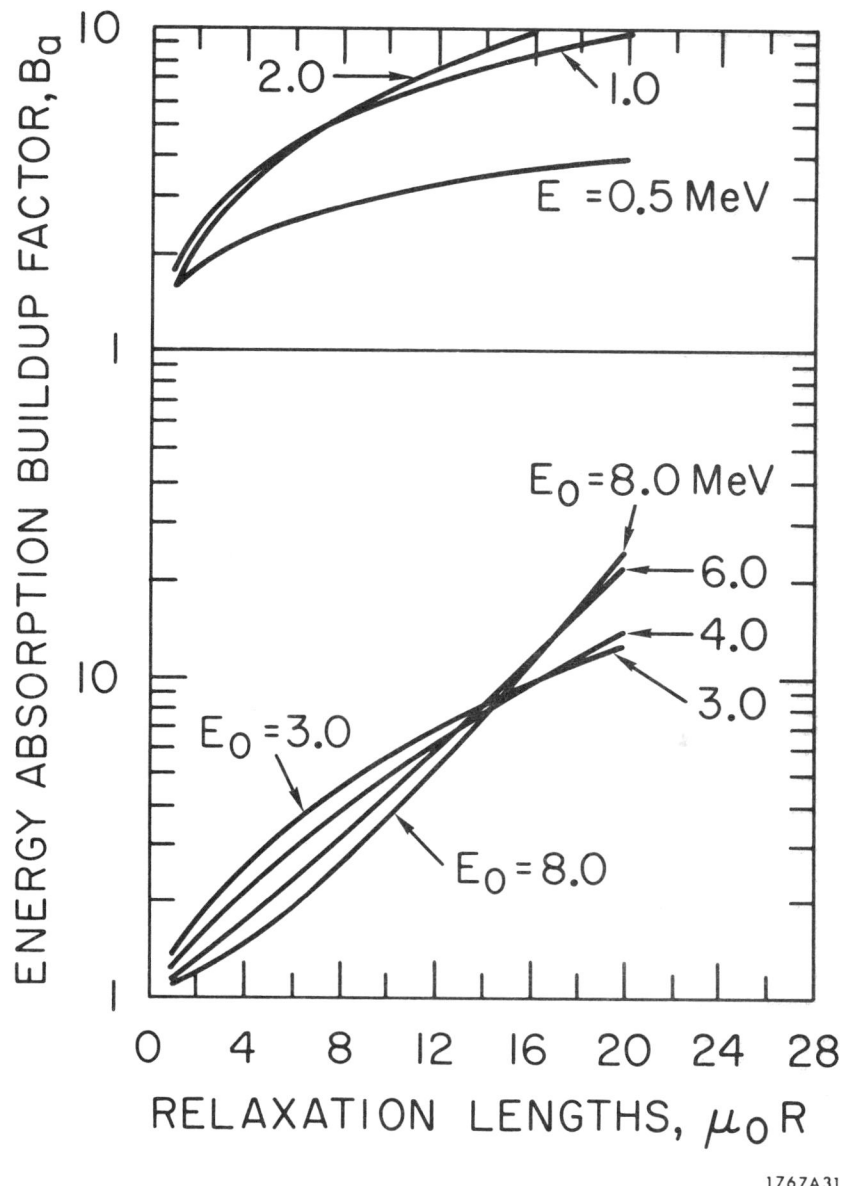

FIG. A.26

Energy absorption buildup factor in lead for a point isotropic source.

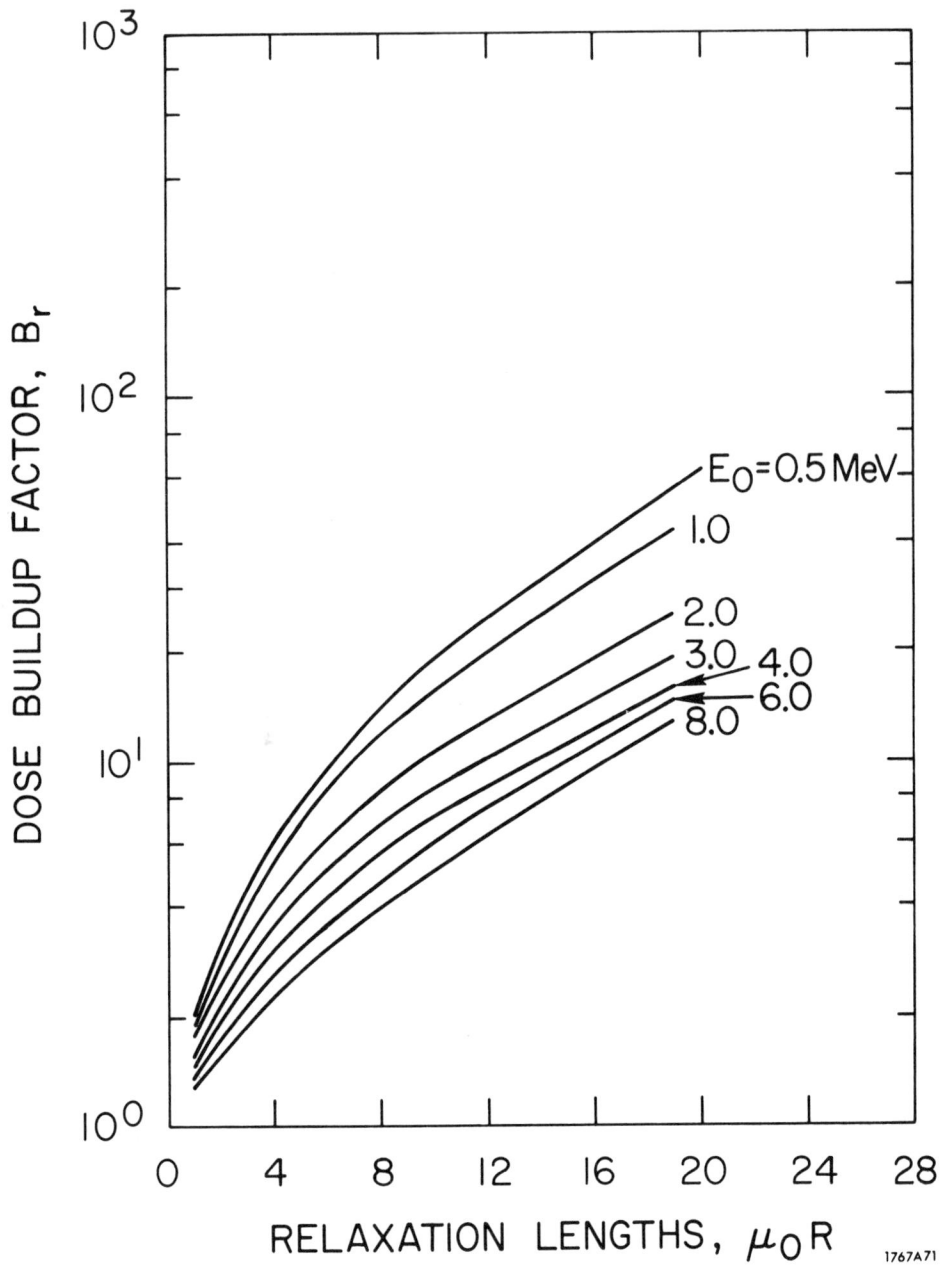

FIG. A.27

Dose buildup factor in iron for a point isotropic source.

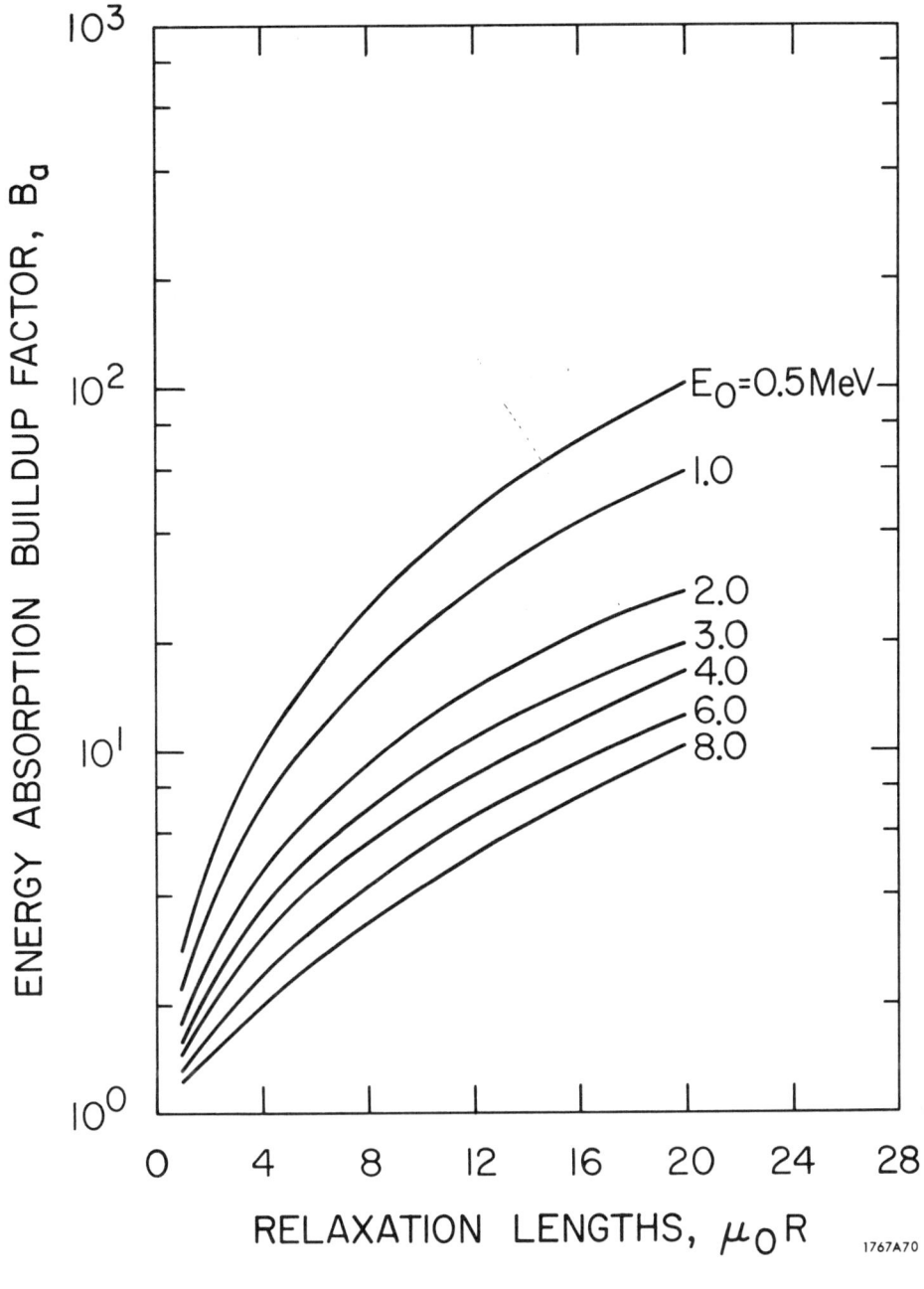

FIG. A.28

Energy absorption buildup factor in iron for a point isotropic source.

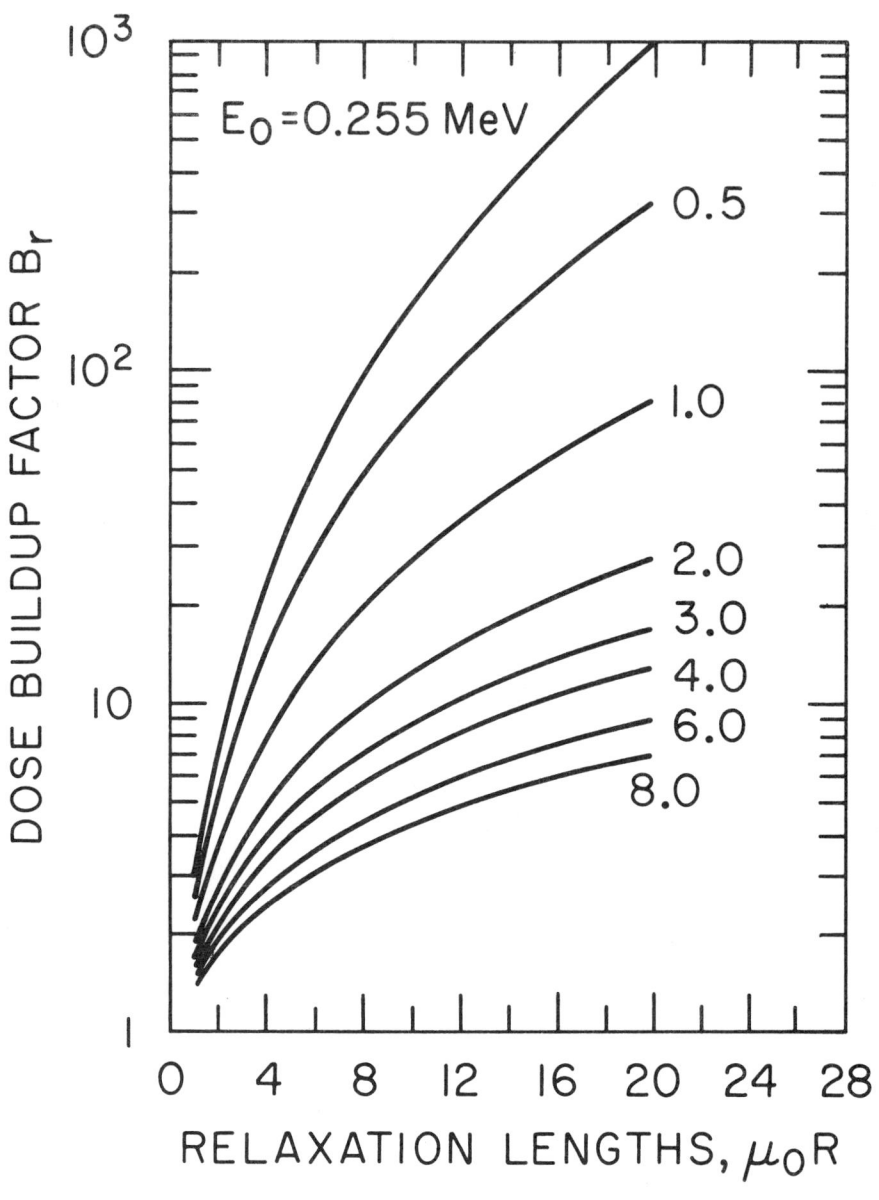

FIG. A.29

Dose buildup factor in water for a point isotropic source.

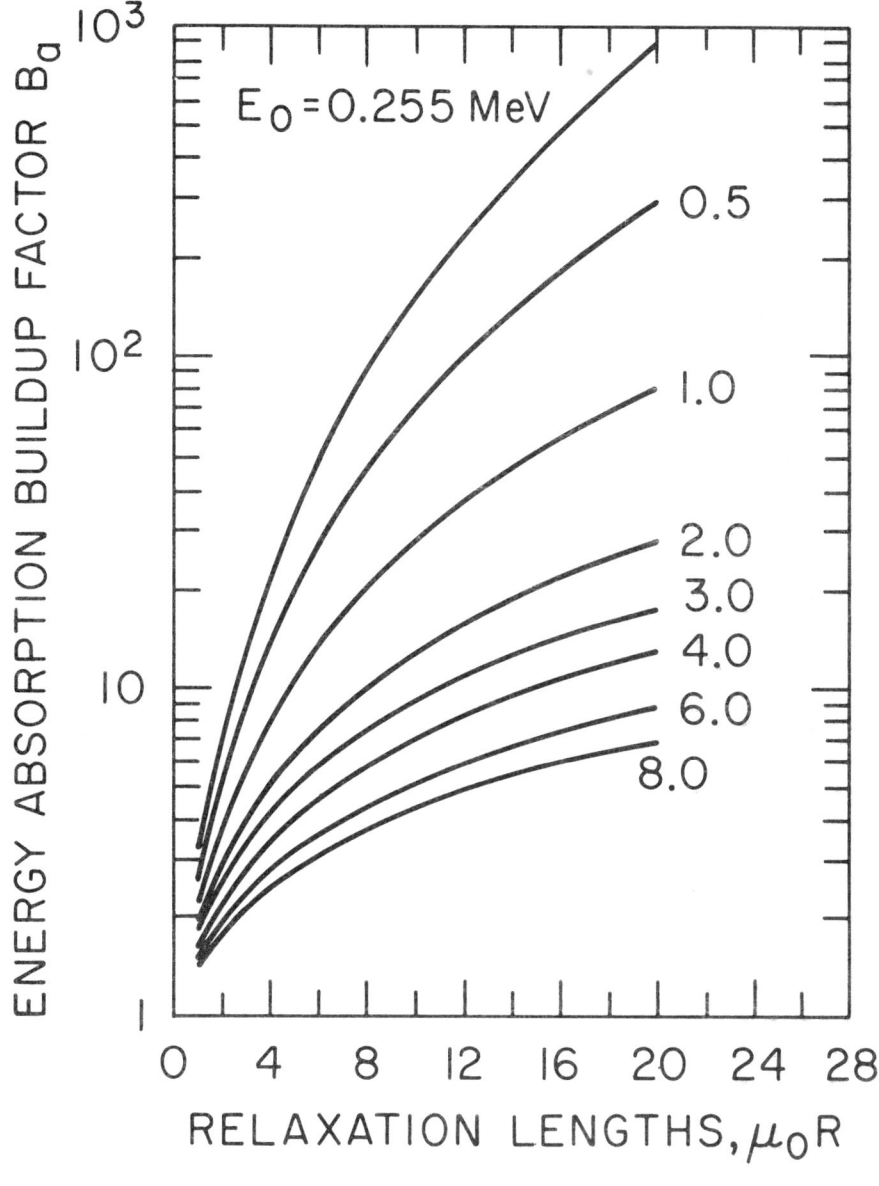

FIG. A.30

Energy absorption buildup factor in water for a point isotropic source.

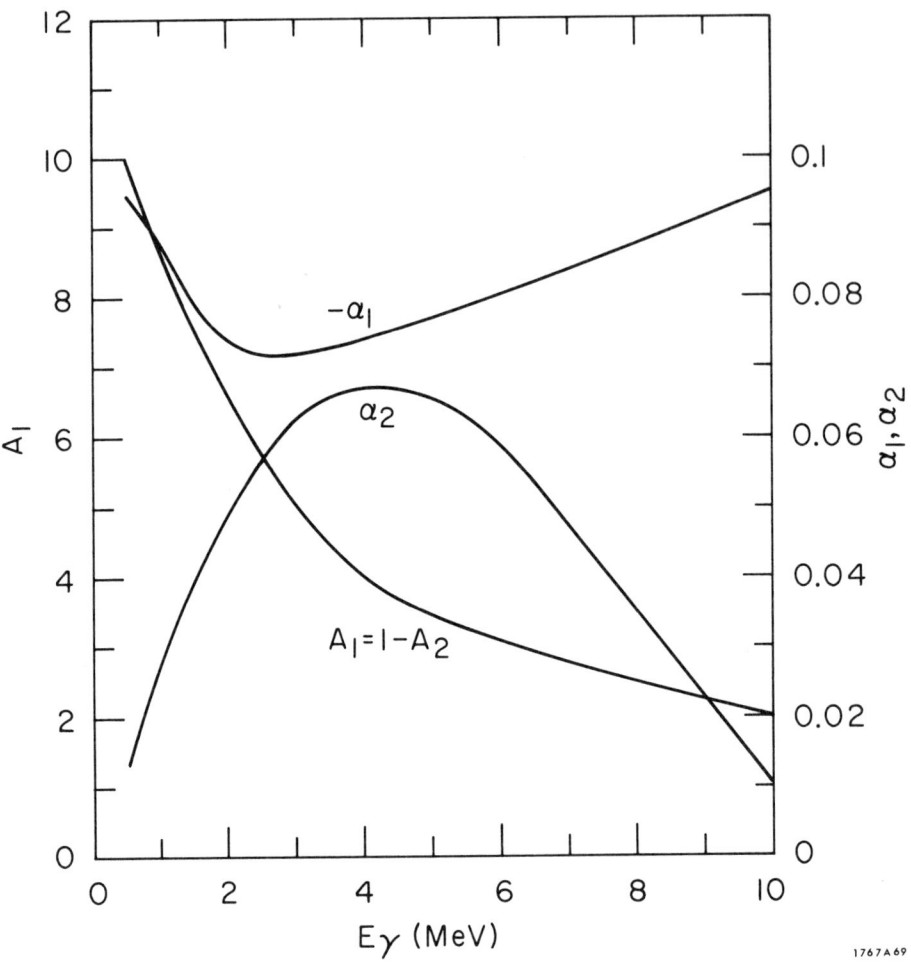

FIG. A.31

Dose buildup factor in iron for a point isotropic source.

$$B = A_1 e^{-\alpha_1 \mu x} + A_2 e^{-\alpha_2 \mu x}$$

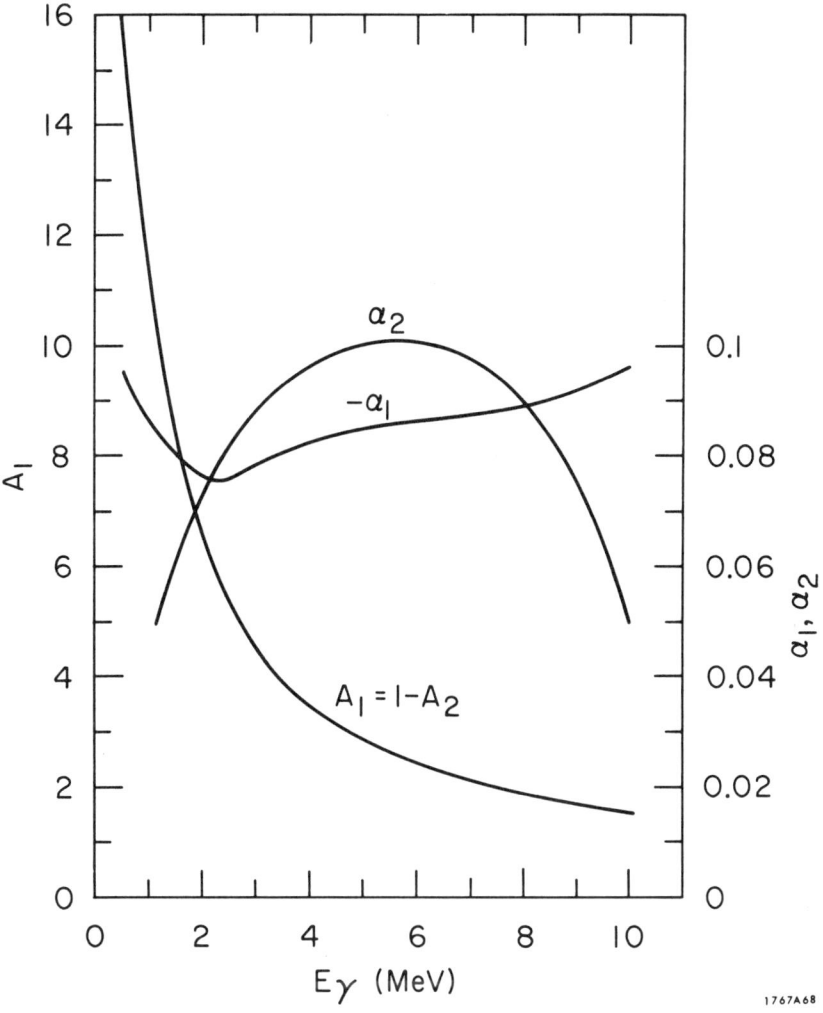

FIG. A.32

Energy absorption buildup factor in iron for a point isotropic source.

$$B = A_1 e^{-\alpha_1 \mu x} + A_2 e^{-\alpha_2 \mu x}$$

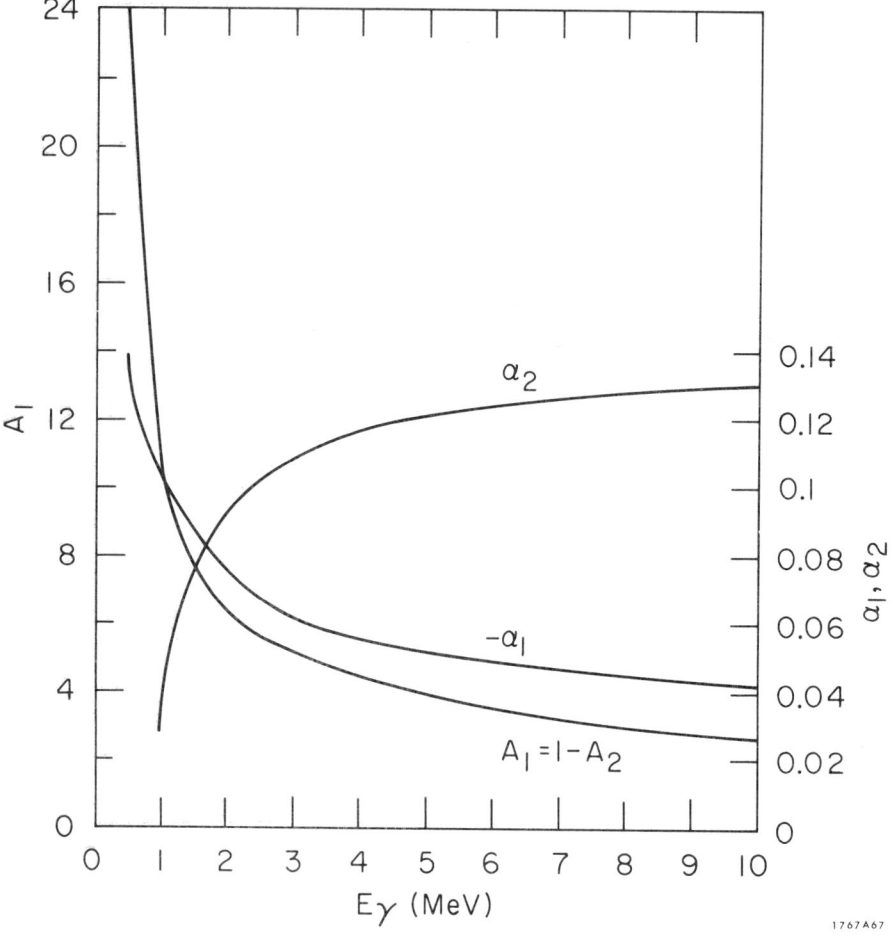

FIG. A.33

Dose buildup factor in water for a point isotropic source.

$$B = A_1 e^{-\alpha_1 \mu x} + A_2 e^{-\alpha_2 \mu x}$$

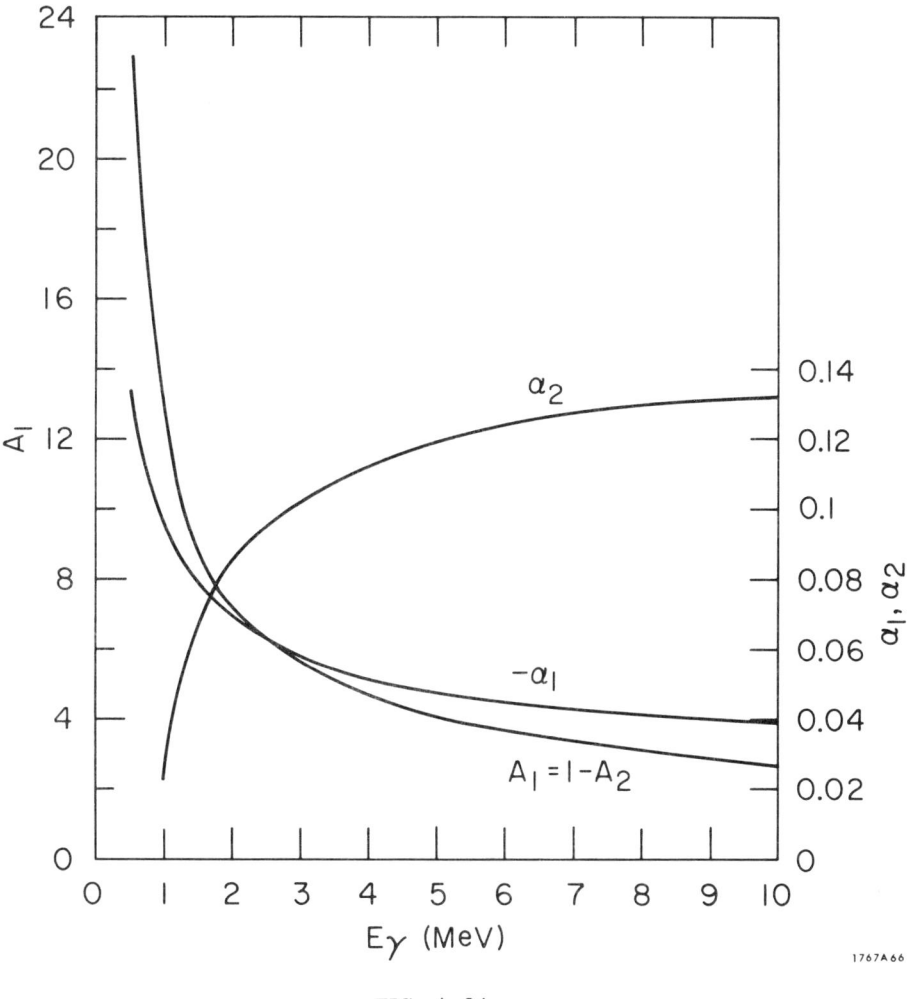

FIG. A.34

Energy absorption buildup factor in water for a point isotropic source.

$$B = A_1 e^{-\alpha_1 \mu x} + A_2 e^{-\alpha_2 \mu x}$$

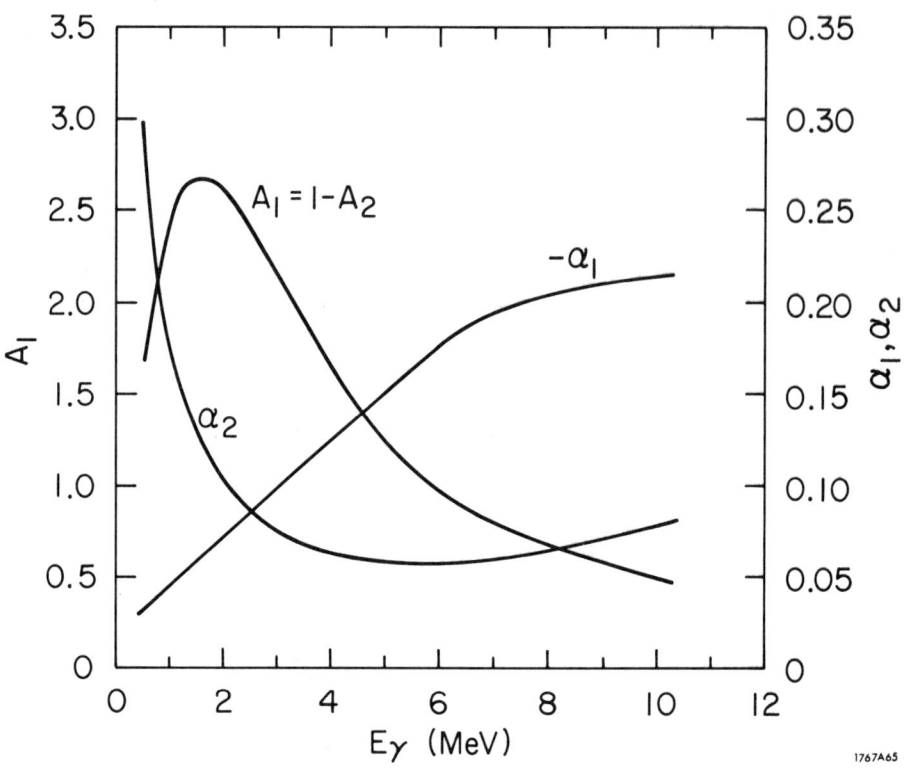

FIG. A.35

Dose buildup factor in lead for a point isotropic source.

$$B = A_1 e^{-\alpha_1 \mu x} + A_2 e^{-\alpha_2 \mu x}$$

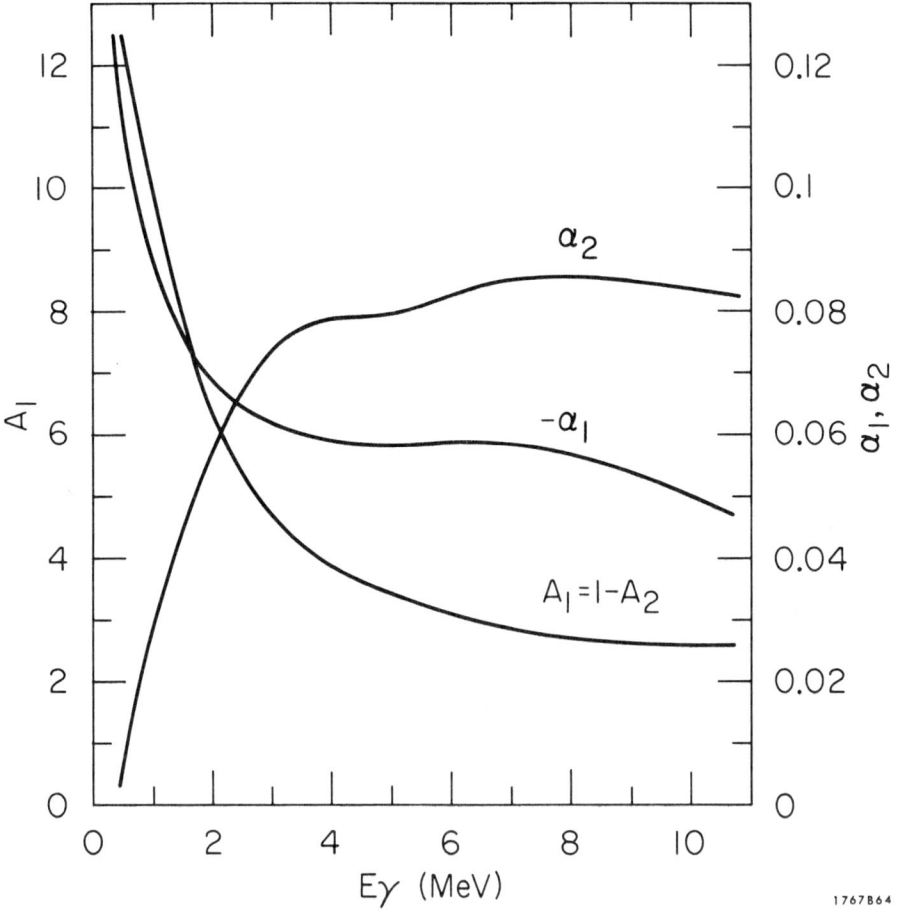

FIG. A.36

Dose buildup factor in concrete for a point isotropic source, $\rho = 2.3$ g/cm^3.

$$B = A_1 e^{-\alpha_1 \mu x} + A_2 e^{-\alpha_2 \mu x}$$

SUBJECT INDEX

A

Absorbed dose   1, 5, 7-12, 85, 91-92, 94, 96-101, 103, 131, 134, 137-143, 145-168

Absorption coefficient   21, 26, 31-32, 131-132

Absorption edge   20-21

Air equivalent   159, 163

Alpha particle (see Charged particles, heavy)

Annihilation radiation   26

Approximation A (of shower theory)   61

Area Source   103, 111-115, 124, 134-135

Attenuation   104-105, 108, 112, 115, 119, 138-139

Attenuation coefficient   3, 30-32, 104-105, 110, 115, 119-120, 124, 129, 135-136, 138-139

Auger electron   21-22

Average energy per ion pair   (see W)

B

Bhabha cross sections   40

Boltzmann transport equation   129-130, 132-134

    method of moments   134

    method of successive scatterings   133-134

    Monte Carlo method   133-134

    straight ahead approximation   132-133

Born approximation   45, 55, 59, 67-72

Bragg's additivity rule   48

Bragg-Gray principle   146-150, 159

Bremsstrahlung   24-26, 34-35, 53-64, 149, 151, 153

Buildup   104, 106, 108, 113, 118, 129, 131-140, 196-207

    multiple layers   136-137

Burch theory   151

Burlin theory   157-158, 162

C

Cascade shower, electron-photon   61, 64, 67, 133

Cavity chambers   7

    cavity and wall material different   153, 157-163

    devices other than ionization chambers   158, 162-163

    matched walls and cavity   153, 158-160, 162-163

    theory   145-168

        Bragg-Gray   146-153, 159

        Burch   146, 151

        Burlin   146, 157-158, 162

        intermediate cavity   146, 155, 157-159

        large cavity   146, 155-159, 163

        Lawrence   146, 150, 153-154

        small cavity   146-159, 163

        Spencer-Attix   146, 151-154, 157

        stopping power ratio   147, 150-156

Cerenkov radiation   45

Characteristic angle, pair production and bremsstrahlung   26

Charged particle equilibrium (CPE)   4, 7-12, 140, 142-143, 145, 149-150, 152, 158, 160

Charged particle interactions   34-84

Charged particles, heavy
- absorbed dose   143
- radiation loss   61-64
- scattering   67-78
- stopping power   44, 89

Coefficients (see Absorption, Attenuation, Energy absorption, Energy transfer coefficients)

Collision
- hard   34, 38-47, 86-89
- kinematics   35-38
- loss (see Energy loss, collision)
- probability   38-43, 86, 88
- soft   34, 43-45, 47, 89
- stopping power (see Stopping power)

Compton effect   14-17, 26-32, 57-58, 129-131, 133

Critical energy   59-60, 66-67

Cross section
- Bhabha   40
- inactivation   93
- Massey-Corbin   40
- Møller   38
- photon   14, 32

Cutoff energy (see also LET)   151, 158

Cylindrical source   103, 118-122

D

Delta rays   47-48, 86-91, 145, 149, 151-153, 158

Density effect (see Polarization effect)

Disc source (see Area source)

Dose (see Absorbed dose, Dose equivalent, Exposure)

Dose equivalent   2

Dose distribution factor   2

E

Electrons

    absorbed dose   143, 157

    attenuation   158

    continuous energy loss model   150-151

    discrete energy loss model   151, 162

    LET   47-48, 89

    radiation loss   58-61

    scattering   68, 73, 75-76

    secondary (spectrum)   145-148, 151, 155, 157, 159, 162

    stopping power   47-48, 152, 159

Energy, most probable   52-53

Energy absorption coefficient   4, 32, 131, 139-142, 156-157, 159-163

Energy density (see Local energy density)

Energy fluence   3, 5, 142

Energy flux, flux density   3, 103, 128, 131, 135-136

Energy imparted to matter   2, 7-11, 138, 148-149, 152

Energy loss

    collision   43-53, 59-61, 153

    most probable   53

    radiative   58-64, 153

Energy transfer coefficient   3, 32

Energy transferred

    charged particles by collision   35-38, 86-91

    photon   7-11

Equilibrium (see Charged particle equilibrium)

Event size   93-94, 98

Excitation   34, 43, 168

Exponential integrals   172-184

Exposure   1, 7, 131, 134, 141-142, 158-167

Extrapolation chamber   157

F

Fano theorem   158-159

Fermi-Eyges multiple scattering theory   76-78

Feynman diagram   25, 61

Fluence (see also Energy fluence)   2, 5, 142, 145

Fluorescent radiation   21-22, 32

Flux, flux density   2, 5, 103-104, 131, 135-136, 139-141

    calculations   103-137, 141

Fluctuations, in energy loss

    collision

        Gaussian   48-53

        Landau   53

    radiation   64

Fluctuations, in range (see Straggling)

Free-air ionization chamber   7

Frequency (see Probability)

G

Gamma rays (see Photons)

Gaussian scattering (see Scattering, charged particle, multiple)

Giant resonance (see also Photo nuclear)   18

Gray (unit)   2

Gray's principle of equivalence   149

I

Impact parameter   41-43, 46, 63, 69, 88

Internal conversion   22

Ionization   34, 43-53, 149-150, 157, 159-160, 163

Ionization chambers   132, 157-162

Ionizing radiation   1

K

Kerma   3, 5, 7-11

Klein-Nishima formula   16, 30, 129

Knock-on (see Collision, hard)

L

Landau distribution   53

Laurence theory   146, 150, 153-154

LET (linear energy transfer)   4, 47-48, 85-101

    cutoff energy   48, 89-91

    distributions   91-93

      dose average   92-93

      number average   93

      track average   92

Line source   103, 106-110, 135-136

Local energy density   5, 85-88, 93-101

M

Mass (e.g., mass attenuation coefficient, see name of the quantity without "mass" prefixed; e.g., attenuation coefficient)

Massey-Corbin cross section   40

Mesons (see Charged particles, heavy)

Moliere scattering   78

Møller cross section   38

Moments method (see Boltzmann transport equation)

Monte Carlo method (see Boltzmann transport equation)

Moseley's law   20

Mott scattering formula

    electrons   68

    heavy particles   68

N

Nuclear photo effect (see Photo nuclear)

Number flux density (see Flux, flux density)

P

Pair production   15-16, 18, 22-28, 32, 63

Photoelectric effect   14-16, 19-22, 32

Photo meson   15, 17

Photons
- absorbed dose   137-143, 145-166
- interactions   14-33
- LET distributions   91-93
- mass absorption coefficient   31-32
- mass attenuation coefficient   3, 30-32, 138-139
- mass energy absorption coefficient   4, 32, 131, 139-142
- mass energy transfer coefficient   3, 32
- sources (see Sources)

Photo nuclear   14-15, 18

Plane source (see Area source)

Point source   103-106, 110, 114-115, 125, 127, 130, 134, 140-141

Polarization effect   45-47, 53, 150, 159

Positrons (see Electrons)

Probability
- of collision   38-43, 86
- of elastic scattering   67-74
- of increment of local energy density   97-101
- of local energy density   97-101
- of radiation energy loss   53-58

Protons (see Charged particles, heavy)

Q

Quality factor   2, 91

R

Rad (unit)   2

Radiation probability   53-58

Radiation length   24, 58-61, 75

Radiation loss (see Energy loss, radiative)

Radiative process (see Bremsstrahlung)

Range   64-67, 78-82, 167

    scaling laws   78-82

RBE (relative biological effectiveness)   2, 93

Relaxation length   134, 137

Rem (unit)   2

Roentgen (unit)   1, 160

Rutherford formula

    elastic scattering distribution   68-72, 74

    energy distribution   41-44

S

Scattering

    charged particle   67-78

        multiple   72, 76-78

        plural   72, 78

        Rutherford   68-72, 74, 78

        single (see Scattering, charged particle, Rutherford)

    coherent   14-16

    Compton   14-17, 26-32, 129-131, 133

    Delbruck   15, 17

    elastic   14-16, 67-78

    elastic nuclear   15-17

    electron resonance (see Scattering, Rayleigh)

    incoherent   14-16

    inelastic   14-16

    mean square angle of   74-76

    nuclear potential (see Scattering, Delbruck)

    nuclear resonance   15, 17

    photon (see also Buildup)   14-18, 26-32, 129, 139

    Rayleigh   15, 18, 32

    Thomson   17

Screening   54-56, 58, 68, 72-73

Shell corrections   46

Shielding (see also Attenuation and Buildup)

    multiple layer   136-137

Shower (see Cascade shower)

Sievert integrals   185-190

Slab source   103, 115-118, 136

Sources   103-128

    distributions   104, 109, 113, 118

    geometry

        point   103-106, 110, 114-115, 125, 127, 130, 134-135, 140-141

        line   103, 106-110, 135-136

        area   103, 111-115, 124, 134-135

        slab   103, 115-118, 136

        cylinder   103, 118-122

        sphere   103, 122-128

    self absorption in   104, 109, 115, 119-120, 124, 191-195

    strength   103, 124

Spencer-Attix theory   151

Spherical source   103, 122-128

Statistical fluctuations   48-53, 64-67

Stopping power (see also Energy loss)   4, 43-53, 78-82, 88, 142-143, 159

    density effect   45-47

    of compounds   48

    LET   47-48, 86-101

    restricted (see also LET)   47-48, 88, 152

    scaling laws   78-82

    shell correction   46

    total   153

Stopping power ratio (see also Cavity chambers)   147, 150-163

Straggling

    energy (see Fluctuations, in energy loss)

    range   64-67

Straight ahead approximation (see Boltzmann transport equation)

Successive scattering, method of (see Boltzmann transport equation)

T

Taylor formula   135

Thermoluminescent dosimeters (TLD)   163

Tissue equivalent   98

Track length   91-92

Transport theory   128-130

Triplet production (pair production in the field of an electron)   15, 18, 23, 25-28

U

Uncollided flux (approximation)   129-133, 135-136

V

Volume source (see Sources)

W

W (energy per ion pair)   2, 142, 149, 159, 168

X

X-rays (see Photons and Bremsstrahlung)